William F. Shanahan

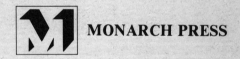 MONARCH PRESS

Essential Math, Science, and Computer Terms for College Freshmen

Copyright © 1981 by William F. Shanahan
All rights reserved
including the right of reproduction
in whole or in part in any form
Published by Monarch Press
A Simon & Schuster Division of
Gulf & Western Corporation
Simon & Schuster Building
1230 Avenue of the Americas
New York, New York 10020

MONARCH PRESS and colophon are trademarks of Simon
& Schuster, registered in the U.S. Patent and Trademark
Office.

Designed by Irving Perkins
Manufactured in the United States of America
10 9 8 7 6 5 4 3 2 1

Library of Congress Catalog Card Number: 79-3323
ISBN 0-671-18435-0

Contents

Preface   7

Introduction   9

Math Terms   13
    Index of Math Terms, 13
    Definitions of Math Terms, 16
    Math Quiz, 99

Science Terms   118
    Index of Science Terms, 118
    Definitions of Science Terms, 122
    Science Quiz, 215

Computer Terms   231
    Index of Computer Terms, 231
    Definitions of Computer Terms, 233
    Computer Quiz, 274

Preface

Familiarity with the vocabulary used in any academic course, business, organization, or profession is absolutely mandatory if people working in those areas expect to succeed. Lack of knowledge can cause failure, anger, frustration, and loss of time and effort.

College instructors will very often assume that students are familiar with the meanings of certain words either because they should have learned the words in prerequisite courses or because the words are relatively standard. Often, however, students have not thoroughly mastered certain words or, because they have seldom used them, have forgotten their meanings. Foreign students studying in the United States for the first time might be familiar with technical words in their own language but have not translated these words into English.

Students with these problems will have difficulty in taking examinations, listening to lectures, or doing homework. Instructors will often not have the time to review the meaning of standard technical words for these students. As a result, many students will undergo unhappy academic experiences.

The use of this publication should help students overcome some of these problems. It may be used as a personal reference book or as a text in a classroom or in a tutorial situation organized to teach or to refresh students who are about to enter college.

Introduction

Objectives

This book was prepared to assist students in or about to enter college as science, mathematics, or engineering majors. The book is intended to serve as:

1. a language reference for foreign students wishing to familiarize themselves with technical English words that they may not have been exposed to in the study of English in their own country.

2. a reference text in English-only classes for foreign students studying in the United States.

3. a refresher text for American students about to enter college.

4. a refresher course for adults returning to school after an extended period of time away from academic work.

5. a supplement in tutorial programs for students having difficulty with technical vocabulary and to prepare students to take college placement examinations.

Preparation and General Description of the Book

The words in this book were taken from the mathematics, science, (physics, chemistry) and computer textbooks used by large universities in freshman courses. The selection was made from those words that instructors would probably assume students had a mastery of from their prior schooling but that are often vague, forgotten, or not understood by students, both American and foreign.

The text is divided into three sections:

1. mathematics

2. science (physics, chemistry)

3. computer science

Each word in a section has a simple description and example of the word. Study of the text should give students a good working knowledge of the words so that they will be familiar with their use in placement tests or during classroom presentations.

A quiz is provided at the end of each section to enable students to test themselves after studying the words in that section.

While there are several definitions for many words, the words in this text have been described in a manner that should provide students with a working knowledge and understanding of the word in the general sense and as it is most often used in college.

An index of the words in each subject is given at the beginning of each section.

Format

Each of the words is separately stated. Under the word, a definition is given in as simple a style as possible. After the definition, at least one example is given to help clarify the use of the word. Many of the examples utilize pictures or sketches to assist the student. The words are arranged alphabetically in each section. An example of the format of the entries follows.

Hypotenuse

DEFINITION:

The side opposite the right angle in a right triangle.

EXAMPLE:

Side c is the hypotenuse of the right triangle.

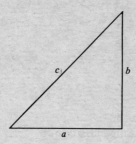

MATH TERMS
Index of Math Terms

Abscissa
Absolute Value
Acute (Angle)
Addition
Adjacent (Side)
Algebraic (Sum)
Algorithm
Altitude
Amplitude
Analytical Geometry
Angle
Antilogarithm (Antilog)
Appendix
Arc Cosine
Arc Sine
Arc Tangent
Area
Associative
Asymptote
Axiom
Axis

Base
Binomial
Binomial Theorem

Cancel
Characteristic
Chord
Circular
Circumference
Coefficient

Colinear
Combination
Commutative
Complement
Complex Number
Component
Cone
Congruent
Conjugates
Constant
Converse
Coordinates
Coplanar
Corollary
Cosecant
Cosine
Cotangent
Coterminal (Angles)
Critical Point
Cube (Figure)
Cube (Number)
Cylinder

Decimal and Decimal Point
Degree
Denominator
Dependent
Dependent Variable
Determinant
Diagonal
Diameter
Difference

14 | Essential Math, Science, and Computer Terms

Directrix
Dividend
Divisor
Domain

Elevation (Angle of Elevation)
           (Angle of Depression)
Ellipse
Equation
Estimate
Exponent
Expression (Math)

Factor
Factorial
Field
Finite
Fluctuate
Formula
Fraction
Function

Graph
Grid (System)
Group

Harmonic Series
Horizontal
Hyperbola
Hypotenuse

Identity
Imaginary (Number)
Independent Variable
Inequality
Infinity
Integral (Sign)
Integration
Interpolation
Intersection
Irrational

Junction

Least Common Denominator
Least Common Multiple
Line
Linear
Linear Function
Locus
Logarithm
Logarithmic
Lowest Terms

Major
Mantissa
Matrix
Maximum
Mean
Median
Minimum
Minor
Mirror Image
Modulus
Monomial
Multiplicative Inverse

Negative
Normal
Numerator

Oblique
Obtuse (Angle)
Ordinate
Origin

Parabola
Parallel
Parallelepiped
Parallelogram
Parameter
Parentheses
Partial Fraction

Math Terms | 15

Peak
Percent
Perimeter
Permutations
Perpendicular
Pivot
Plane
Point
Polar Coordinates
Polynomial
Positive
Postulate
Power
Product
Progression
Proof
Proportion
Pythagorean Theorem

Quadrant
Quadratic Equation
Quadrilateral
Quotient

Radian
Radical
Radical Sign
Radicand
Radius
Range
Rate
Ratio
Rational Number
Real Number
Reciprocal or Multiplicative Inverse
Reducible
Reflection
Reflexive (Axiom of Equality)
Relation
Remainder
Revolution (Rotation)

Rhombus
Root
Root Mean Square (R.M.S.)

Scalar
Scientific Notation
Secant
Set
Simplify
Sine
Slope
Solution
Sphere
Spiral
Square
Square Root
Statistics
Substitution (Axiom of Equality)
Subtend
Subtraction
Sum
Supplement
Symbol
Symmetric (Axiom of Equality)
Symmetry

Table
Tangent
Terms of an Equation
Theorem
Transitive (Axiom of Equality)
Translation of Axes
Triangle
Trigonometric

Value
Variable
Vector
Vertex
Vertical
Volume

Definitions of Math Terms

Abscissa

DEFINITIONS:

1. The horizontal distance on a graph from a point to the y axis
2. The first coordinate in an ordered pair of numbers

EXAMPLES:

1. The *abscissa* of point P

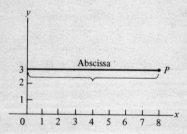

2. If $P = (8, 3)$, them 8 is the *abscissa* of P

Absolute Value

DEFINITION:

For any nonzero real number b, the *absolute value* of b is the larger of the numbers b and $-b$ ($-b$ in this case means the opposite of b). An *absolute value* is indicated by $|\ |$.

EXAMPLES:

To find $|10|$, compare 10 and -10, then choose the larger. So $|10| = 10$.

$-1| = 1$

$|-1| = 1$ $\qquad |-\tfrac{1}{2}| = \tfrac{1}{2}$

Acute (Angle)

DEFINITION:

An angle of less than 90 degrees

EXAMPLES:

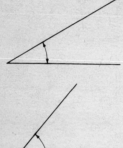

is an *acute* (less than 90°) *angle*.

is an *acute* (less than 90°) *angle*.

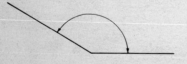

is NOT an *acute angle*. It is greater than 90°.

Addition

DEFINITION:

The arithmetic process of finding the total value of two or more values. The answer is called the sum. The *addition* process uses the sign +.

18 | **Essential Math, Science, and Computer Terms**

EXAMPLES:

1. $\begin{array}{r} 421 \\ + 312 \\ \hline 733 \end{array}$ Sum

2. $\frac{1}{4} + \frac{1}{2} + \frac{3}{8} = \frac{9}{8}$ Sum

Adjacent (Side)

DEFINITION:

Being next to something else.

EXAMPLES:

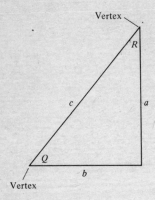

a. Side *b* is *adjacent* to angle *Q*.

b. Side *a* is *adjacent* to angle *R*.

c. Side *a* is *adjacent* to side *c* and side *b*.

Algebraic (Sum)

DEFINITION:

The final value of a number of expressions after they have been added together. This result takes into consideration the signs of the numbers being summed up.

EXAMPLE:

$-7 + 5 + 6 - 2 = 2$

2 is the *algebraic sum* of the above expression.

Algorithm

DEFINITION:

A systematic approach or means that enables a person to solve a problem in a certain number of steps, frequently by repeating an operation.

EXAMPLE:

A long-division problem

$$
\begin{array}{r}
1\,1\,1\,1\,1 \\
3\,{\overline{\smash{\big)}\,3\,3\,3\,3\,3}} \\
\underline{-3} \\
3 \\
\underline{-3} \\
3 \\
\underline{-3} \\
3 \\
\underline{-3} \\
3 \\
\underline{-3} \\
0
\end{array}
$$

In this example, certain systematic steps are performed in a particular order when solving the problem.

Altitude

DEFINITION:

The length of the perpendicular from a point to a given line.

EXAMPLE:

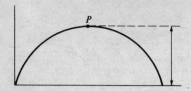

The *altitude* of point *P*
(can also be called amplitude)

Amplitude

DEFINITION:

The maximum height of an oscillating wave above the mean value (given as a positive number)

EXAMPLE:

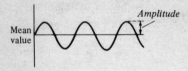

Analytical Geometry

DEFINITION:

Geometry using algebra and coordinates to define lines or graphs in space

EXAMPLE:

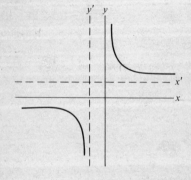

Example of a hyperbola located on a graph whose coordinates have been shifted (dotted lines)

Angle

DEFINITION:

A set of points formed by the union of two distinct non-colinear rays with a common end point

EXAMPLES:

1.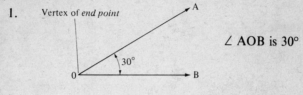

   $\angle$ AOB is 30°

2.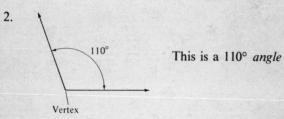

   This is a 110° *angle*

Antilogarithm (Antilog)

DEFINITION:

The number that is represented by a logarithm. If $\log_b x = a$, then x is the antilog of a.

EXAMPLE:

100 is the *antilogarithm* of 2 in $\log_{10} 100 = 2$.

Appendix

DEFINITION:

A part of a book, usually located at the end, that lists various types of information related to portions of the book proper

EXAMPLES:

1. *Appendix* 1 of a math book might be tables of squares and square roots.

2. *Appendix* 2 of a math book might be tables of logarithmic values.

Arc Cosine

DEFINITION:

The angle whose cosine is

EXAMPLES:

1. If $y = $ cosine x, then x is the *arc cosine* of y.

2. If .5 = cosine 60°, then 60° is the *arc cosine* of .5.

Arc Sine

DEFINITION:

The angle whose sine is

EXAMPLES:

1. If $y = $ sine x, then x is the *arc sine* of y.

2. If .5 = sine 30°, then 30° is the *arc sine* of .5.

Arc Tangent

DEFINITION:

The angle whose tangent is

EXAMPLES:

1. If $y = $ tangent x, then x is the *arc tangent* of y.

2. If 1 = tangent 45°, then 45° is the *arc tangent* of 1.

Area

DEFINITION:

A measure of a surface bounded by closed lines or curves or both; measured in square units such as square feet or square yards

EXAMPLE:

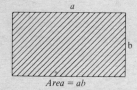
Area = ab

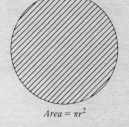
Area = πr^2

Shaded portions indicate the *area*.

Associative

DEFINITION:

Pertaining to a mathematical operation that gives an equal expression when terms are grouped without a change in their order

EXAMPLES:

1. $(a + b + c) = (a + b) + c$
2. $(a + b) + c = a + (b + c)$
3. $(a)(bc) = (ab)(c)$

Asymptote

DEFINITION:

A line that approaches a curve but never reaches it

EXAMPLE:

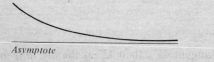
Asymptote

Axiom

DEFINITION:

Something accepted as correct without a proof

EXAMPLE:

If x and y are different real numbers, then either:
x is greater then y, or
y is greater then x.

(Our common sense tells us that the above statement has to be correct.)

Axis

DEFINITION:

A reference line indicating a direction on a graph

EXAMPLE:

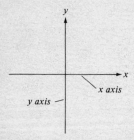

Base

DEFINITION:

The number that underlies a numerical system

EXAMPLES:

1. In $147_{10} = 1(10)^2 + (10)^1 + 7$, the *base* is 10.
2. The *base* of $\log_{10} x$ system is 10.
3. If $10^n = x$, 10 is the *base*.

Binomial

DEFINITION:

The indicated sum or difference of two values

EXAMPLES:

1. $c^4 + 3a$.
2. $a^4 - 7b$.

Binomial Theorem

DEFINITION:

The expansion of an expression in the form of $(x+y)^n$ where *n* is a positive integer

EXAMPLES:

1. $(x + y)^2 = x^2 + 2xy + y^2$
2. $(x + y)(x^2 + 2xy + y^2) = x^3 + 3x^2y + 3xy^2 + y^3$

Cancel

DEFINITION:

To take away by division

EXAMPLE:

$$\frac{1}{\cancel{2}} \times \frac{\cancel{2}}{8} = \frac{1}{8}$$

The 2 in the denominator *cancels* out the 2 in the numerator.

Characteristic

DEFINITION:

The integer part of a logarithm

EXAMPLES:

1. If log a = 1.6771,
   1 is the *characteristic*.

2. If log b = 42.8371,
   42 is the *characteristic*.

Chord

DEFINITION:

A line segment that intersects two different points on a curve or circle

EXAMPLES:

1.

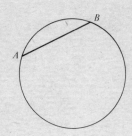

Line segment $\overline{AB}$ is a *chord* on this circle.

2.

Line segment $\overline{CD}$ is a *chord* on this curve.

Circular

DEFINITION:

Having the form or taking the form of circle; round

EXAMPLE:

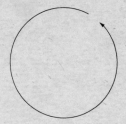

The form is *circular* in shape.

Circumference

DEFINITION:

A measure of the perimeter or outer boundary of a circle; the "length" of the circle

EXAMPLE:

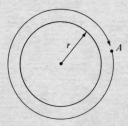

A measure of the length from point A all the way around the circle back to A.

$$Circumference = C = 2\pi r$$

Coefficient

DEFINITION:

Any factor of a product is called the *coefficient* of the product of the remaining factors

EXAMPLES:

In $5y^3x$:

5 is the numerical *coefficient* of y^3x.

y^3 is the *coefficient* of $5x$.

x is the *coefficient* of $5y^3$.

Colinear

DEFINITION:

In the same straight line

EXAMPLES:

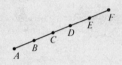

Points *A, B, C, D, E,* and *F* are *colinear*.

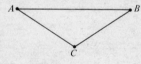

Points *A, B,* and *C* are NOT *colinear* since they are all not in the same straight line.

Combination

DEFINITION:

A number of items arranged into a group

EXAMPLE:

To obtain a *combination* of $4 = r$ numbers from $6 = N$ given numbers, use the following formula:

$$_NC_r = \frac{N!}{r!(N-r)!} = \frac{6 \times 5 \times 4 \times 3 \times 2 \times 1}{4 \times 3 \times 2 \times 1\,(2 \times 1)} = 15$$

Commutative

DEFINITION:

Pertaining to a mathematical operation in which one term operating with another is equal to the second term operating with the first

EXAMPLES:

1. $x + y = y + x$
2. $a + b = b + a$

Complement

DEFINITION:

When the sum of two angles equals 90°, one angle is said to *complement* the other

EXAMPLE:

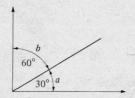

Angle a is the *complement* of angle b. (Angle b is also the *complement* of angle a.)

Complex Number

DEFINITION:

A number that has two parts, one that is real and one that is imaginary. The number can be shown in the form $a + bi$ where a and b are real numbers and i is equal to $\sqrt{-1}$.

EXAMPLE:

$4 + 7i$

Component

DEFINITION:

A part of a whole

EXAMPLES:

1. $a + b = c$
   a and b are *components* of the above equation.

2. In the ordered pair (6,19), 6 is the first *component* and 19 the second *component*.

Cone

DEFINITION:

A solid figure obtained by revolving a right triangle about one of its legs

EXAMPLE:

This is a right circular *cone*. Sets of points obtained by cutting a plane through a *cone* are called conic sections.

Examples of conics are:

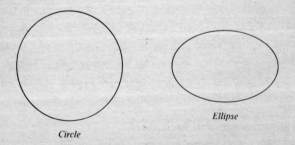

Circle   Ellipse

Math Terms | **31**

Parabola   Hyperbola

Congruent

DEFINITION:

Equal or agreeing in all ways

EXAMPLES:

1. $\dfrac{1}{2} = \dfrac{2}{4} = \dfrac{3}{6} = \dfrac{4}{8}$ All of the fractions are *congruent* since they are equal (when the numbers are divided, they all produce the same answer).

2. 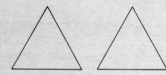 The two triangles are *congruent* since one can be placed exactly over the other.

Conjugates

DEFINITION:

Numbers that have the form $a + bi$ and $a - bi$ (when a and b are real) are *conjugates* of each other

EXAMPLES:

1. $1 + i\sqrt{11}$ and $1 - i\sqrt{11}$
 are *conjugates* of each other.
2. $-2 + 1\sqrt{5}$ and $-2 - i\sqrt{5}$
 are *conjugates* of each other.

Constant

DEFINITION:

A quantity that does not change

EXAMPLES:

1. e always equals 2.71828 and is a *constant*.

2. π is always the same value and is a *constant*.

Converse

DEFINITION:

The exchanging of a statement and its conclusion

EXAMPLES:

1. *Statement*: If $x = 10$, then $3x + 4 = 34$.
   Converse of the above statement:
   If $3x + 4 = 34$, then $x = 10$.

2. *Statement*: Equal chords of a circle are the same distance from the center of the circle.
   Converse: Chords that are the same distance from the center of a circle are equal.

Coordinates

DEFINITION:

The values used to indicate a point in a fixed system such as the coordinate plane

EXAMPLE:

The *coordinates* of point P are (4,3).

Coplanar

DEFINITION:

In the same plane

EXAMPLE:

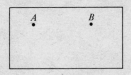

Points A and B are *coplanar*.

Corollary

DEFINITION:

A fact that is proved while something else is being proved

EXAMPLE:

If it was proved that a^2 is greater than zero, then a *corollary* would be that a is a real number not equal to zero.

Cosecant

DEFINITION:

In a right triangle, the ratio between the hypotenuse and the side opposite an angle

EXAMPLE:

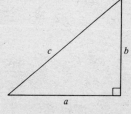

$$\text{Cosecant } \theta = \frac{c}{b}$$

Note: *Cosecant* is the reciprocal of the sine; $\text{sine } \theta = \frac{b}{c}$

Cosine

DEFINITION:

In a right triangle, the ratio between the side next to an angle (other than the 90° angle) and the hypotenuse

EXAMPLE:

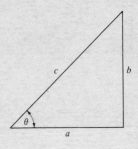

$$\text{Cosine } \theta = \frac{a}{c}$$

Note: Secant is the reciprocal of the *cosine*; secant $\theta = \frac{c}{a}$

Cotangent

DEFINITION:

In a right triangle, the ratio between the side next to an angle and the side opposite that angle

EXAMPLE:

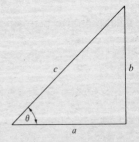

$$\text{Cotangent } \theta = \frac{a}{b}$$

Note: *Cotangent* is the reciprocal of the tangent; tangent $\theta = \frac{b}{a}$

Coterminal (Angles)

DEFINITION:

Angles that have the same two sides

EXAMPLE:

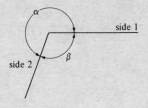

Angle α and β are *coterminal* since the same sides make up the angles for both.

Critical Point

DEFINITION:

A point $(X_1 Y_2)$ on the graph of the smooth function $f(x)$, such that $f'(x_1) = 0$

EXAMPLE:

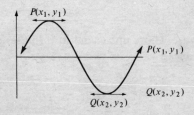

P and Q are *critical points* of the graph of $f(x)$.

Cube (Figure)

DEFINITION:

A rectangular parallelepiped (a right rectangular prism) all of whose edges are equal

EXAMPLE:

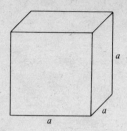

Cube   (Number)

DEFINITION:

A number or expression taken as a factor three times; an expression raised to the third power

EXAMPLES:

$3^3 = 3 \times 3 \times 3$

$(4 + x)^3 = (4 + x)(4 + x)(4 + x)$

$5 \times 5 \times 5 = 5^3$

Cylinder

DEFINITION:

A solid figure whose upper and lower bases are circles in parallel planes

EXAMPLE:

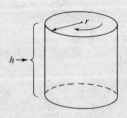

A right circular *cylinder*
Volume = $\pi r^2 h$

Surface area = $2\pi r(h + r)$

Math Terms | 37

Decimal and Decimal Point

DEFINITIONS:

1. Pertaining to tenths or to the number 10
2. A dot that indicates tenths of a number

EXAMPLES:

1. *Decimal* .3 = 3/10
2. *Decimal* .128 = 128/1,000
3. *Decimal point* is the dot .

Degree

DEFINITION:

A unit of measure of an angle or arc of a circle

EXAMPLE:

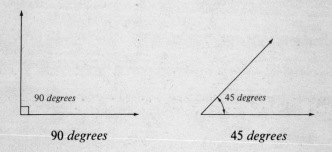

90 *degrees*   45 *degrees*

Represented by ° as in 30°

180° Arc

Denominator

DEFINITION:

The number in a fraction that is below the fraction bar (the line)

EXAMPLE:

In $\frac{48}{6} = 8$; 6 is the *denominator*.

Dependent

DEFINITIONS:

1. Influenced or determined by something else
2. Having its value determined by the value of another

EXAMPLE:

$3x = y$

The value of y is *dependent* on the value given x.

Dependent Variable

DEFINITION:

A symbol whose value is determined by the value of another symbol

EXAMPLE:

$2x + x^2 = y$

y is the *dependent variable* because its value depends on the value assigned to x.

Determinant

DEFINITION:

A way of getting answers to certain algebraic equations by writing the equation differently (in a square array)

EXAMPLE:

The determinant of $x_1y_2 - x_2y_1$ is written:

x_1y_1
x_2y_1

Diagonal

DEFINITION:

A line segment that joins two nonconsecutive vertices of a polygon

EXAMPLES:

1.

   Line a is a *diagonal* of this quadrilateral.

2.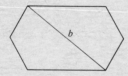

   Line b is a *diagonal* of this polygon.

Diameter

DEFINITION:

A line segment joining two points on a circle (chord) that passes through the center of the circle

EXAMPLE:

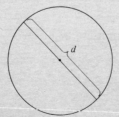

Difference

DEFINITIONS:

1. The amount by which one value is greater than another
2. The answer obtained in a subtraction problem in arithmetic

EXAMPLES:

1. $$\begin{array}{r} 176 \\ -83 \\ \hline 93 \end{array} \quad \textit{Difference}$$

   The *difference* between 176 and 83 is a value of 93.

2. $$\begin{array}{r} 47 \\ -46 \\ \hline 1 \end{array} \quad \textit{Difference}$$

   The *difference* between 47 and 46 is a value of 1.

Directrix

DEFINITION:

A line that is stationary (fixed) and used to help define a curve

EXAMPLES:

See examples under **Ellipse, Hyperbola** and **Parabola.**

Dividend

DEFINITION:

A number or quantity to be divided by another

EXAMPLES:

1. $\frac{100}{25} = 4$

   100 is the *dividend.*

2. $\frac{33}{11} = 3$

33 is the *dividend*.

Divisor

DEFINITION:

The number by which the dividend is divided to produce the quotient

EXAMPLES:

1. $100 \div 5 = 20$

    5 is the *divisor*, 100 is the dividend, and 20 is the quotient.

2. $\frac{75}{3} = 25$

    3 is the *divisor*, 75 is the dividend, and 25 is the quotient.

Domain

DEFINITION:

The set of all first components or first coordinates of a relation

EXAMPLE:

Given the relation [(0,1), (2,3), (8,10)], the set [0,2,8] is the *domain* of the relation.

Elevation (Angle of Elevation) (Angle of Depression)

DEFINITION:

Elevation: The angle formed by a horizontal ray and the line of sight from the observer's eye to a higher point

Depression: The angle formed by a horizontal ray and the line of sight from the observer's eye to a lower point

EXAMPLE:

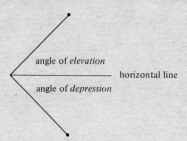

Ellipse

DEFINITION:

The locus of a point that moves so that the sum of its distances from two fixed points is a constant

EXAMPLE:

Ellipse

Equation

DEFINITION:

A relationship between known or unknown terms that are linked by an equal sign

EXAMPLES:

1. $3x + x = 12$
2. $x^2 + 2xy + y^2 = 42$
3. $x = y$
4. $1 + 1 = 2$

Math Terms | 43

Estimate

DEFINITION:

1. A guess about a numerical value
2. To guess the numerical value

EXAMPLES:

1. The newspaper's *estimate* was that about 60,000 people saw the parade.
2. The mayor's advisers *estimate* that unemployment will decrease by 10%.

Exponent

DEFINITION:

The number that indicates the power to which a quantity (base) is raised

EXAMPLES:

1. In x^6, where 6 is the *exponent*, $x^6 = x \cdot x \cdot x \cdot x \cdot x \cdot x$ (x is taken as a factor 6 times)
2. In $\left(\frac{1}{2}\right)^3$; where 3 is the *exponent*, $\left(\frac{1}{2}\right)^3 = \left(\frac{1}{2}\right)\left(\frac{1}{2}\right)\left(\frac{1}{2}\right)$.

Expression (Math)

DEFINITION:

The showing of a value in the form of symbols

EXAMPLES:

1. $1 + 2$, $\frac{18}{6}$, 3×1 are all numerical *expressions* of the number 3.
2. Two more than one-third of x can be written as the variable *expression* $\left(2 + \frac{1}{3} x\right)$.

Factor

DEFINITION:

Any numbers or quantities that multiplied together give a result

EXAMPLES:

1. 3 and 4 are *factors* of 12.
2. 2 and 6 and *factors* of 12.
3. 12 and 1 are *factors* of 12.
4. 3 and 3 are *factors* of 9.
5. $(x + 1)$ and $(x - 1)$ are *factors* of $x^2 - 1$.

Factorial

DEFINITION:

The product of a positive number and all the positive whole numbers below it down to 1

EXAMPLE:

Factorial $7 = 7 \times 6 \times 5 \times 4 \times 3 \times 2 \times 1$

Note: *Factorial* is written !, so in the above example $7! = 7 \times 6 \times 5 \times 4 \times 3 \times 2 \times 1$.

Field

DEFINITION:

An abstract mathematical system given by the following:

    elements: $a, b, c, \ldots$ etc., of a set P
    operations: $+, \times$
    axioms:

EXAMPLES:

1. closure $(a + b \in P \,;\, a \times b \in P)$

Math Terms | 45

2. associative law $(a + b) + c = a + (b + c)$ and $(a \times b) \times c = a \times (b \times c)$
3. zero element $(a + 0 = 0 + a = a)$
4. unit element $(a \times 1 = 1 \times a = a)$
5. additive inverse $a + (-a) = (-a) + a = 0$
6. cummutative law $a + b = b + a$ and $a \times b = b \times a$
7. multiplicative inverse $a \times \frac{1}{a} = \frac{1}{a} \times a$
8. distributive law $a \times (b + c) = (a \times b) + (a \times c)$

Finite

DEFINITION:

Capable of being reached or completed by counting

EXAMPLE:

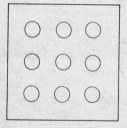

There are a *finite* number of circles in the box.

Fluctuate

DEFINITION:

To constantly change back and forth

EXAMPLES:

1.

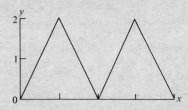

The points in this graph are *fluctuating* between $y = 2$ and the x axis.

2.

This curve is *fluctuating* around the y axis.

Formula

DEFINITION:

A way of showing the equal relationship between certain quantities

EXAMPLES:

1. $I = \dfrac{E}{R}$ is the *formula* showing in symbolic form that the current I is equal to the voltage E over the resistance R.

2. $F = ma$ is the *formula* showing in symbolic form that force F is equal to the mass m times the acceleration a.

Fraction

DEFINITION:

A part of a whole integer or a part of a unit

EXAMPLES:

1. $\frac{1}{4}$

2. $\frac{4}{5}$

3. $\frac{7.1}{8.5}$

4. $\frac{293}{2}$

Function

DEFINITION:

A relation (a set of ordered pairs) in which each first component corresponds exactly to a second component

EXAMPLE:

In $A = [(0,1), (1,4); (2,3), (5,8)]$, A is a *function*.
In $B = [(0,1), (1,4); (2,3), (1,12)]$, B is not a *function*.

In A, the difference between (0,1) and (1,4) is 1 and 3, same as the difference between (2,3) and (5,8).

In B, the difference between (0,1) and (1,4) is not the same as the difference between (2,3) and (1,12).

Graph

DEFINITION:

A picture representation of two or more variables showing how a change in one reflects a change in another

48 | **Essential Math, Science, and Computer Terms**

EXAMPLE:

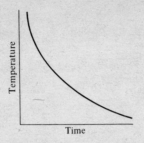

A *graph* showing the relationship between temperature and time

Grid (System)

DEFINITION:

A system of horizontal and vertical lines that can be used as a reference for locating points, lines, curves, etc.

EXAMPLE:

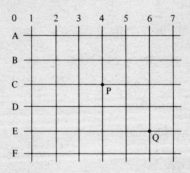

A *grid* system that locates point *P* at location C4 and point *Q* at E6

Group

DEFINITION:

To place elements within a boundary

Math Terms | 49

EXAMPLE:

In $a + b + c = (a + b + c)$,

the values of *a*, *b*, and *c* on the left of the equation have been *grouped* into the parentheses ( ) on the right side of the equation.

Harmonic Series

DEFINITION:

A series of terms in which the reciprocals* of the terms are in arithmetic progression

EXAMPLE:

$$1 + \frac{1}{2} + \frac{1}{3} + \frac{1}{4} + \frac{1}{5} + \frac{1}{6} + \ldots$$

*$\frac{1}{2}$ is the reciprocal of $\frac{2}{1}$; $\frac{4}{5}$ is the reciprocal of $\frac{5}{4}$

Horizontal

DEFINITION:

Parallel to the horizon; at a right angle to the vertical

EXAMPLE:

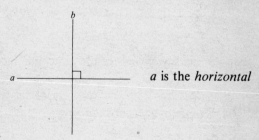

a is the *horizontal*

Hyperbola

DEFINITION:

In a plane, a *hyperbola* is the set of all points located at a distance from a fixed point (called a focus) that is equal to its distance from a fixed line (called a directrix) multiplied by a constant greater than 1

EXAMPLE:

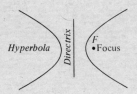

Hypotenuse:

DEFINITION:

The side opposite the right angle in a right triangle

EXAMPLE:

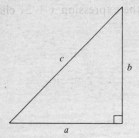

Side c is the *hypotenuse* of the right triangle.

Identity

DEFINITION:

An equation that is true for all permitted values of the variables

EXAMPLE:

$4x \equiv 3x + 1x$

$\equiv$ is the sign for identity.

CHAPTER 1

1644 WELTON · DENVER, COLO. 80202

	F 2	***007.95
	F 2	***001.95
S F		***009.90
	FCLTX	***000.65
S F		***010.55

10-77-0047-ooo385 TOTAL F CASH ***010.55

THANK YOU

Imaginary   (Number)

DEFINITION:

    A number that includes the square root of a negative real number

EXAMPLE:

    1. $4\sqrt{-1}$ is an *imaginary* number because $\sqrt{-1}$ is part of it.

    2. $\sqrt{-1}$ is sometimes denoted by *i*, so $6i$ is also an *imaginary* number.

Independent Variable

    A symbol whose value determines the value of other symbols

EXAMPLE:

$f(x) = x + 2x$

    x is the *independent variable* because as it is assigned different values, the value of the expression $x + 2x$ changes.

    In the above example:

        1. $f(1) = 3$
        2. $f(4) = 12$

Inequality

DEFINITION:

    An expression of this form: $a \neq b$, where $\neq$ means "not equal to"

EXAMPLES:

    1. $2 \neq 3$;

    2. $(a + b) \neq a - b$

Infinity

DEFINITION:

    A value that is greater than any other assigned value

EXAMPLE:

The sign for *infinity* is ∞.

Integral (Sign)

DEFINITION:

A symbol used in mathematics, generally calculus, that indicates you wish to determine the area under the curve represented by the expression following the ∫ sign

EXAMPLE:

In $2x^2 + 2 + 1\ dx,$

∫ is the *integral* sign and you wish to determine the area under the curve $2x^2 + 2 + 1$ between the values of $x = 0$ and $x = 4$.

Integration

DEFINITION:

A method used to find a function when its derivative is given. By using it, you may find the area of a figure and the volume of a solid of revolution, among other things.

EXAMPLES:

4 $\times dx$

∫ $\sin^3 \times dx$

Interpolation

DEFINITION:

The process of filling in values between other known values

EXAMPLE:

Given known values of the cosine of 9° and the cosine of 10°, find the value of the cosine of 9° 30′ by *interpolation*.

Angle	Cosine
9°	.98770
10°	.98480
	.00290

Difference between cosine of 9° and cosine of 10°

9° 30′ is halfway between 9° and 10°.

One-half the difference between cosine 9° and cosine 10°:

$$.00290 \times \frac{1}{2} = .00145$$

.98770   cosine 9°
.00145   less one-half the difference between cosine 9° and cosine 10°
.98625   cosine 9° 30′.

By interpolation:

Angle	Cosine
9°	.98770
9° 30′	.98625
10°	.98480

Intersection

DEFINITION:

1. The *intersection* of two sets is the set of all elements common to both sets

2. The points where lines or curves or both cross each other

EXAMPLES:

1.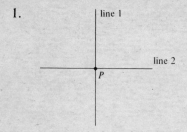

   Point *P* is the *intersection* of lines 1 and 2.

2.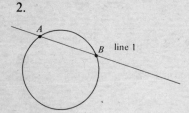

   A and *B* are the points of *intersection* of line 1 and the circle.

3.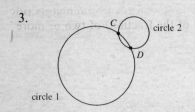

   C and *D* are the points of *intersection* of circle 1 and circle 2.

4. If $A = [1, 2, 3, 12]$
   $B = [-5, 2, 11, 12, 99]$
   Then $A \cap B$ (read *A* intersects *B*) = $[2, 12]$.

Irrational

DEFINITION:

Not capable of being shown as an integer or the quotient of two integers

EXAMPLES:

1. $\sqrt[3]{5}$ $\sqrt[-4]{3}$ $\sqrt[7]{64}$ are *irrational* numbers.

2. $x^2 = 2$ is *irrational* since the equation has no rational roots.

Junction

DEFINITION:

The point or place at which two or more objects or lines meet

EXAMPLE:

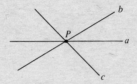

Point *P* is the *junction* of lines *a*, *b*, and *c*.

Least Common Denominator

DEFINITION:

The smallest common multiple of the denominators (the numbers below the line of the fraction) of two or more fractions

EXAMPLES:

1. In $\frac{1}{8} + \frac{1}{4} + \frac{1}{16}$, 16 is the *least common* denominator since it is the smallest or lowest number into which 16, 8, and 4 (all the denominators) will divide.

2. In $\frac{1}{8} + \frac{1}{7}$, 56 is the *least common denominator*.

Least Common Multiple

DEFINITION:

The smallest positive whole number that is exactly divisible by two or more whole numbers

EXAMPLE:

1. The *least common multiple* of 5, 6 and 10 is 30.

2. The *least common multiple* of 4, 6, and 8 is 24.

Line

DEFINITION:

The path of a moving point; it has length but no width

EXAMPLE:

A —————— B   A *line* between points A and B.

Linear

DEFINITION:

Along a line (not circular), in one direction only

EXAMPLES:

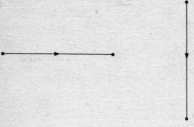

Linear Function

DEFINITION:

An expression of the form $y = mx + b$ where m and b are real numbers and m is not equal to zero.

EXAMPLES:

1. $y = -3x + 7$
2. $f(x) = \dfrac{5}{2} x - 8$

Locus

DEFINITION:

The path of a point as it moves according to some equation or law

EXAMPLES:

1.

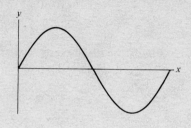

Part of the *locus* of all the points of a sine wave.

2.

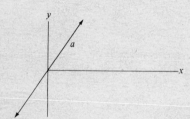

Line *a* is the *locus* of the points of the equation $x = y$.

Logarithm

DEFINITION:

The number to which a given base must be raised to yield a specific result.

EXAMPLES:

$\log_{10} \frac{1}{10} = -1$ since $10^{-1} = \frac{1}{10}$

$\log_2 8 = 3$ since $2^3 = 8$

Logarithmic

DEFINITION:

Pertaining to a measurement scale wherein an increase one unit results in the quantity measured being multiplied by 10

58 | Essential Math, Science, and Computer Terms

EXAMPLES:

If $10^x = y$, then as x increases by 1, y increases as follows:
$10^1 = 10$
$10^2 = 100$
$10^3 = 1,000$

As the exponent increases by 1, the result increases tenfold.

The log of 10 is 1.
The log of 100 is 2.
The log of 1,000 is 3.

Lowest Terms

DEFINITION:

A fraction is in *lowest terms* if the largest integer that can divide both numerator and denominator is 1

EXAMPLES:

1. $\frac{2}{3}$ is in *lowest terms*.

2. $\frac{5}{10}$ is NOT in *lowest terms*. Divide numerator and denominator by 5, and $\frac{5}{10} = \frac{1}{2}$.

$\frac{1}{2}$ is in *lowest terms*.

1. In + + , 16 is the *least common denominator* since it is the smallest or lowest number into which 16, 8 and 4 (all the denominators) will divide.

Major

DEFINITION:

Larger in size, amount, or number

EXAMPLE:

The *major* number of the following series is 942.

1, 12, 161, 712, 942

Mantissa

DEFINITION:

That part of a logarithm after the decimal point

EXAMPLES:

1. If $\log_b a = 1.6771$, .6771 is the *mantissa*.
2. If $\log_c d = 3.8013$, .8013 is the *mantissa*.

Matrix

DEFINITION:

A set of numbers arranged in columns and/or rows, usually so that linear equations may be solved

EXAMPLES:

1.

$$\text{rows} \begin{bmatrix} \overbrace{5 \quad 4 \quad 8 \quad 12}^{\text{columns}} \\ 1 \quad 2 \quad -6 \quad 4 \\ 3 \quad -7 \quad 9 \quad 11 \end{bmatrix}$$

2. $\begin{bmatrix} 1 \\ 9 \end{bmatrix}$ a column *matrix*

3. $\begin{bmatrix} 1 & * & 7 \end{bmatrix}$ a row *matrix*

Maximum

DEFINITION:

The highest point on a graph or the largest value

EXAMPLES:

1.

This point is the *maximum* point of the curve.

2. 1,000, 100, 10
   1,000 is the *maximum* of the above three values.

Mean

DEFINITION:

Generally, the average of several values

EXAMPLE:

The *mean* of 100, 200, 300, 400, and 500 = $\frac{1{,}500}{5}$ = 300

Median

DEFINITION:

1. In a triangle, a line segment that joins a vertex with the midpoint of the opposite side
2. The middle number in a set of numbers ordered according to magnitude

EXAMPLES:

1.

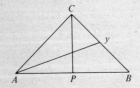

Line *CP* is the *median* line *AB*.
Line *AY* is the *median* to line *CB*.

2. 1, 2, 3, 4, 5, 6, 7, 8, 9, 10, 11
   6 is the *median* in this sequence.

Minimum

DEFINITION:

The smallest value or lowest point on a graph

Math Terms | 61

EXAMPLES:

1.

This point is in the *minimum* of the curve.

2. 1,000, 100, 10
10 is the *minimum* of the above three values.

Minor

DEFINITION:

The smallest in size, amount, or number

EXAMPLE:

The minor number of the following series is 1,000.
1,000, 2,000, 3,000, 4,000

Mirror Image

DEFINITION:

The picture of an object that you see when it is reflected in a mirror

EXAMPLE:

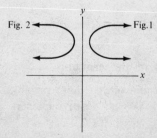

Fig. 2 is the *mirror image* of Fig. 1; consider the y axis to be the mirror.

Modulus

DEFINITIONS:

1. For a complex number, the *modulus* is computed by adding the squares of each part of the complex number and taking the positive square root of the sum
2. A certain value that gives the same remainder when it is the divisor of two quantities
3. The factor by which the logarithm of one base is multiplied to change it to a logarithm of another base

EXAMPLES:

1. The *modulus* of $a + bi$ is $\sqrt{a^2 + b^2}$
2. $a = b$ (*modulus* m) is the same as $a - b = km$ for some integer k.
   For example, $17 = 7$ (*modulus* 5) because $17 - 7$ is divisible by 5 for the integer $k = 2$.
3. $\log_{10} 64 = 1.8062$
   $\log_2 64 = 6$

   Since the log of 64 to the base 10 is 1.8062, the *modulus* of 64 in converting its log to the base 2 is 3.32 — that is, $(1.8062)(3.32) = 6$.

Monomial

DEFINITIONS:

1. Consisting of one term only
2. A number and one or more variables whose exponents of the variables are positive integers

EXAMPLES:

1. $3x$
2. $5x^2y^3$

Math Terms | 63

Multiplicative Inverse

DEFINITION:

A number that when multiplied by the given number yields 1 as a result

EXAMPLES:

1. $\frac{1}{4}$ is the *multiplicative inverse* of $\frac{4}{1}$ since $\frac{1}{4} \cdot \frac{4}{1} = 1$.

2. $\frac{\frac{1}{2}}{\frac{1}{4}}$ is the *multiplicative inverse* of $\frac{\frac{1}{4}}{\frac{1}{2}}$

Negative

DEFINITION:

A sign or value in the minus direction

EXAMPLES:

1. −7

2.

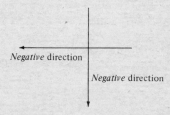

Normal

DEFINITION:

A line that makes a 90° angle to a surface or line.
Normal can be a substitute word for perpendicular.

EXAMPLES:

1.

ray 1

A ─────── B

Ray 1 is *normal* to line AB.

2. 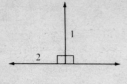 Rays 1 and 2 are *normal* to each other.

3.  Line *l* is *normal* to the arc at point *P*.

Numerator

DEFINITION:

The number in a fraction that is above the fraction bar (the line)

EXAMPLE:

In $\frac{14}{7} = 2$, 14 is the *numerator*

Oblique

DEFINITION:

A line that is slanting or declining, not perpendicular or parallel

EXAMPLES:

1.  Ray 2 is *oblique* to ray 1.

2.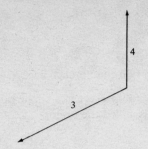

Ray 3 is *oblique* to ray 4.

Obtuse (Angle)

DEFINITION:

> An angle that is greater than 90 degrees but less than 180 degrees

EXAMPLE:

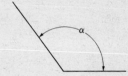

> The angle is *obtuse* because it is greater than 90° but less than 180°.

Ordinate

DEFINITION:

> 1. The vertical distance on a graph from a point to the *x* axis
>
> 2. The second coordinate in an ordered pair of numbers

EXAMPLES:

1.

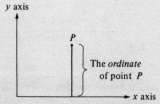

2. If $P = (5,9)$, then 9 is the *ordinate* of P.

Origin

DEFINITION:

1. The point labeled (0,0) on the number line
2. When two or more axes intersect in a single point, the point of intersection is called the *origin*.

EXAMPLE:

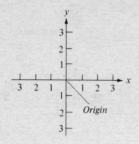

Parabola

DEFINITION:

In a plane, a curve made by moving a point in such a way that it is always equally distant from a fixed point (called a focus) and a fixed line (called a directrix)

EXAMPLE:

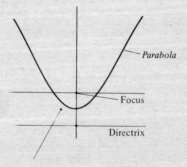

This point is as far from the focus as it is from the directrix. This relationship holds for every other point on the *parabola*.

Parallel

DEFINITION:

1. *Parallel* lines are coplanar lines that do not intersect.
2. *Parallel* planes are planes that do not intersect.

EXAMPLES:

1. 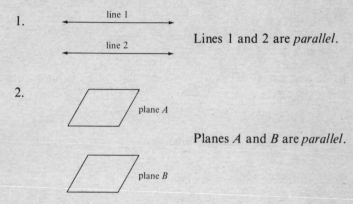 Lines 1 and 2 are *parallel*.

2. Planes *A* and *B* are *parallel*.

Parallelepiped

DEFINITION:

A figure (prism*) with six faces all of which are parallelograms*

EXAMPLE:

*See definitions in this book

Parallelogram

DEFINITION:

A four-sided figure whose opposite sides are parallel

EXAMPLE:

$$\overline{AB} \parallel \overline{CD}$$
and $\overline{BC} \parallel \overline{AD}$

$\parallel$ is the symbol for "is parallel to."

Parameter

DEFINITION:

Keeping an arbitrary constant while other variables are changed

EXAMPLE:

1. $x = 4$ (in all equations)
$x = y + z$

By keeping $x = 4$ and changing $y = 2$, we get:
Let $y = 2$, then $z = 2$
Let $y = 3$, then $z = 1$
Let $y = 4$, then $z = 0$, etc.

In this case, we are changing the *parameters* y and z while keeping x constant.

Parentheses

DEFINITION:

The two curved lines used to enclose mathematical symbols that are to be treated as a single term. One of these curved lines is called a *parenthesis*.

EXAMPLES:

1. ( )
The two curved lines of *parentheses*.

2. $(a + b + c)$
The *parentheses* enclosing the values a, b and c, which when added together are to be taken as one value.

Partial Fraction

DEFINITION:

A part of a whole or a part of something

EXAMPLES:

A fraction can be separated into *partial fractions*. The sum of the *partial fractions* equals the original fraction.

1. $\dfrac{1}{(x^2-1)} = \left(\dfrac{1}{x+1}\right)\left(\dfrac{1}{x-1}\right)$

    $\dfrac{1}{(x+1)}$ and $\dfrac{1}{(x-1)}$ are the *partial fractions*.

2. $\dfrac{1}{2} = \dfrac{1}{4} + \dfrac{1}{4}$

    The $\dfrac{1}{4}$'s are the *partial fractions*.

Peak

DEFINITION:

The high point of a quantity or function

EXAMPLE:

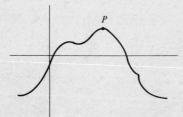

P is the *peak* of this function.

Percent

DEFINITION:

A part of one hundred; a proportion per hundred

EXAMPLES:

1. 40 out of 100 is $\frac{40}{100} \times 100$, or 40 *percent* (40%).

2. 127 out of 462 is $\frac{127}{462} \times 100$, or 27.5 *percent* (27.5%).

Perimeter

DEFINITION:

A measure of the total outer boundary of a geometric figure that is closed

EXAMPLES:

1.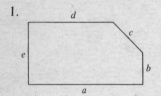

The *perimeter* of this figure is $a + b + c + d + e$.

2.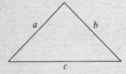

The *perimeter* of this triangle is $a + b + c$.

Permutations

DEFINITION:

Different arrangements of a fixed number of different items

EXAMPLE:

The six *permutations* of the digits 4, 5, and 6 are:

456	564
465	654
546	645

Math Terms | 71

Perpendicular

DEFINITION:

At right angles to each other

EXAMPLES:

1.  Lines *a* and *b* are *perpendicular* to each other.

2. 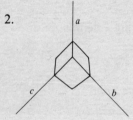 In a solid, lines *a*, *b*, and *c* are all *perpendicular* to each other.

Pivot

DEFINITION:

A fixed point around which something rotates

EXAMPLE:

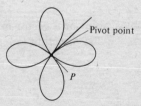

Point *P* is the pivot on which the propeller spins.

Plane

DEFINITION:

A surface such that a straight line joining two of its points lies entirely in the surface

EXAMPLES:

1.

A *plane*

2. The face of this page is a representation of a *plane*.

Point

DEFINITIONS:

1. A small mark or dot
2. An element in geometry having a definite position but no size or shape

EXAMPLES:

1. • The mark to the left is a *point*.
2.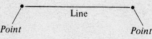
A line between two *points*.

Polar Coordinates

DEFINITION:

A system whereby a point or a vector may be located by knowing the distance and direction (angle) of the point from an origin

EXAMPLE:

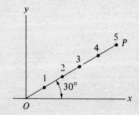

Point *P* at the end of the vector is located by knowing its distance from the origin (5 units) and its direction (30° counterclockwise from the *x* axis).

Polynomial

DEFINITION:

An expression or function that is the indicated sum of two or more terms

EXAMPLE:

$x^2 - 2xy + y^2$

Positive

DEFINITION:

A sign or value in the plus direction

EXAMPLES:

1. $+6$

2.
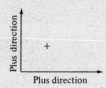

Postulate

DEFINITION:

To assume something is true without proof

EXAMPLE:

If 3 is greater than 1, one can *postulate* that 4 is greater than 1 without having to prove it.

Power

DEFINITION:

A number or value that is multiplied by itself two or more times has been raised to a *power*. The number of times the value is taken as a factor determines its *power*.

EXAMPLE:

5^4 is 5 raised to the 4th *power*.

Product

DEFINITION:

When 2 or more values are multiplied together, you get their *product*

EXAMPLES:

1. $2 \times 2 = 4$; 4 is the *product*.

2. $3 \times 6 = 18$; 18 is the *product*.

3. $111 \times 10 = 1{,}110$; 1,110 is the *product*.

Progression

DEFINITION:

A series of numbers that get larger or smaller by a constant value between numbers

EXAMPLES:

1. 1, 2, 3, 4, 5, 6 is a *progression* of numbers increasing by the value of 1.

2. $1, \frac{7}{8}, \frac{6}{8}, \frac{5}{8}, \frac{4}{8}, \frac{3}{8}, \frac{2}{8}, \frac{1}{8}$ is a *progression* of fractions decreasing by the value of $\frac{1}{8}$.

Proof

DEFINITION:

A way of checking an answer in a subtraction arithmetic problem by using addition

EXAMPLE:

In the problem: $176 - 83 = 93$
Subtraction process

You may prove your answer by adding 93 and 83 to see if it equals 176:

Proof 93 + 83 = 176
Addition process

Proportion

DEFINITION:

An equality of two ratios

EXAMPLES:

1. $x/y = z/v$ is an example of a *proportion*.

2. $x:y = z:v$ is another method of expressing the above *proportion*.

Pythagorean Theorem

DEFINITION:

The theorem that shows the relationship between the sides of a right triangle, i.e., $a^2 + b^2 = c^2$. c is the length of the hypotenuse; a and b are the lengths of the remaining sides

EXAMPLE:

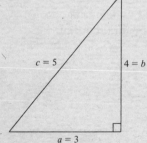

$$a^2 + b^2 = c^2$$
$$3^2 + 4^2 = c^2$$
$$9 + 16 = c^2$$
$$25 = c^2$$
$$5 = c$$

Quadrant

DEFINITION:

One-fourth of a circle or plane

EXAMPLE:

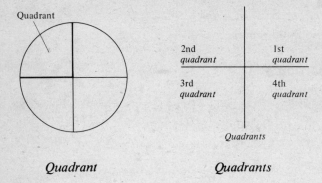

Quadrant *Quadrants*

Quadratic Equation

DEFINITION:

An equation that can be represented in the form $ax^2 + bx + c = 0$, where a, b, and c are elements of [real numbers] and $a \neq 0$

EXAMPLE:

$x^2 - 2x - 35 = 0$
Where $a = 1$
$b = -2$
$c = -35$ may be solved by factoring $(x - 7)(x + 5) = 0$ or by using the quadratic formula:

$$x = \frac{-b \pm \sqrt{b^2 - 4ac}}{2a}$$

Quadrilateral

DEFINITION:

A plane figure determined by 4 distinct points no 3 of which are colinear. If the points are named A, B, C, D, then the segments AB, BC, CD, and DA intersect only at their end points.

EXAMPLES:

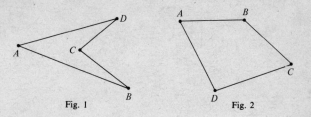

Figs. 1 and 2 are *quadrilaterals*.

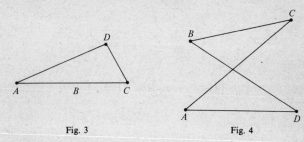

Figs. 3 and 4 are NOT *quadrilaterals*.

Quotient

DEFINITION:

The answer to a division problem

EXAMPLES:

1. $14 \div 7 = 2$ 2 is the *quotient*.
2. $18 \div 6 = 3$ 3 is the *quotient*.

Radian

DEFINITION:

An angle has a measure of one radian if and only if it is a central angle (its vertex is at the center of a circle) and the arc that cuts it is equal to the radius of the circle.

EXAMPLE:

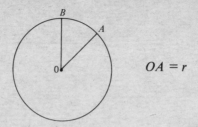

The measured arc AB is equal to r. The angle BOA is 1 radian = 57.3°.

Radical

DEFINITION:

A mathematical expression that appears with a radical sign ($\sqrt{}$)

EXAMPLES:

1. $\sqrt{63}$
2. $\sqrt{x+1}$

Radical Sign

DEFINITION:

A symbol that indicates the root of the number under the sign is to be obtained

EXAMPLES:

1. The figure $\sqrt{}$ is called the radical sign.
2. $\sqrt[3]{9}$ This expression is requesting the cube (3rd) root of 9.

Radicand

DEFINITION:

An expression whose root is to be taken

EXAMPLES:

1. $\sqrt{121} = 11$

2. 121 is the *radicand* (11 is the positive 2nd root or principal square root of 121).

Radius

DEFINITION:

A line segment joining the center of a circle to a point on the circle

EXAMPLE:

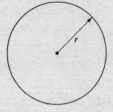

Range

DEFINITION:

The set of all second components or second coordinates of a relation

EXAMPLE:

Given the relation [(0,1), (2,3), (8,10)], the set [1,3,10] is the *range* of the relation.

Rate

DEFINITION:

A quantity of one term measured per unit of another item

EXAMPLES:

1. Fifteen miles per hour, 15 miles/hour, is the *rate* at which a car may be traveling.

2. In some families, the *rate* of taxation may be 50% (50¢ tax per dollar of income).

Ratio

DEFINITION:

The relation one value bears to another; an indicated quotient

EXAMPLES:

1. 100 lbs. and 40 lbs. have the *ratio* 10:4, or 5:2.
2. 16 chickens and 40 ducks have the *ratio* 16:40, or 2:5.

Rational Number

DEFINITION:

A number that may be represented by the quotient $\frac{a}{b}$ where a and b are integers and $b \neq 0$. The set of rational numbers does not include values such as π or roots of negative numbers such as $\sqrt{-3}$

EXAMPLE:

$$-5 \quad -4 \quad -3 \quad -2 \quad -1 \quad 0 \quad 1 \quad 2 \quad 3 \quad 4 \quad 5 \quad 6$$

Rational numbers

Other *rational numbers*: $-\frac{1}{6}$, $\sqrt{16}$, 0, $-\sqrt{25}$

Real Number

DEFINITION:

The set of all positive and negative numbers and zero

EXAMPLES:

$1, -12, \frac{1}{4}, e, \pi, -7.66$.

Reciprocal or Multiplicative Inverse

DEFINITION:

Two numbers are *reciprocals* or *multiplicative inverses* of each other if their product is 1

EXAMPLES:

1. $\frac{1}{5}$ is the *reciprocal* of 5; $\left(\frac{1}{5}\right)(5) = 1$.

2. $\sqrt{3}$ is the *reciprocal* of $1/\sqrt{3} = 1$; $(\sqrt{3})(1/\sqrt{3}) = 1$.

3. $-\frac{3}{4}$ is the *reciprocal* of $-\frac{4}{3}$; $\left(-\frac{3}{4}\right)\left(-\frac{4}{3}\right) = 1$.

Reducible

DEFINITION:

Capable of being changed to lower terms

EXAMPLES:

1. $\frac{12}{20}$ is *reducible* to $\frac{3}{5}$.

2. $\frac{2x}{12}$ is *reducible* to $\frac{1x}{6}$, or $\frac{x}{6}$

3. $\frac{(x+4)}{(x+4)^5} = \frac{1}{(x+4)^4}$

Reflection

DEFINITION:

The turning path of a ray or vector from a surface or plane in another direction

EXAMPLE:

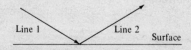

Line 2 is a *reflection* of line 1 off the surface.

Reflexive (Axiom of Equality)

DEFINITION:

Taking two values as equal

EXAMPLES:

1. $a = a$
2. $y = y$
3. $5 = 5$

Relation

DEFINITION:

A set of ordered pairs of numbers

EXAMPLE:

[(0,1), (2,3), (8,10)] is a *relation*.

Remainder

DEFINITION:

A value left over in a division problem after the divisor has been placed into the dividend as many full times as possible

EXAMPLE:

$\frac{23}{7} = 3\frac{2}{7}$; 2 is the *remainder*.

Revolution (Rotation)

DEFINITION:

The motion of a point or line or curve or surface around a point or axis

EXAMPLE:

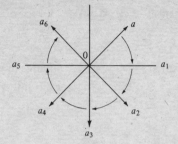

Ray *a* is revolving clockwise around the origin. It is finally located in position a_6.

Rhombus

DEFINITION:

A parallelogram (a four-sided figure whose opposite sides are parallel) with sides of equal lengths

EXAMPLE:

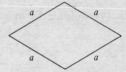

Root

DEFINITIONS:

1. A quantity that multiplied by itself a certain number of times produces a given quantity

2. A number that when substituted for the unknown quantity in an equation will satisfy the equation

EXAMPLES:

1. 3 is the cube *root* of 27: $\sqrt[3]{27} = 3$ since $3 \times 3 \times 3 = 27$.
2. 2 is the fourth *root* of 16: $\sqrt[4]{16} = 2$ since $2 \times 2 \times 2 \times 2 = 16$.
3. In the equation $(x + 1)(x + 2) = 6$, a root is 1.

Root Mean Square (R.M.S.)

DEFINITION:

$$\text{Root mean square} = \sqrt{\frac{\text{sum of squares of individual values}}{\text{total number of values}}}$$

EXAMPLE:

Find the R.M.S. of 1, 2, 3, 4, 5.

$$\text{R.M.S.} = \sqrt{\frac{1^2 + 2^2 + 3^2 + 4^2 + 5^2}{5}} = \sqrt{\frac{1 + 4 + 9 + 16 + 25}{5}}$$

$$= \sqrt{\frac{55}{5}} = \sqrt{11}$$

Scalar

DEFINITION:

A quantity that can be defined by magnitude alone

EXAMPLES:

The *scalar* quantity of the above figures is 5.

Scientific Notation

DEFINITION:

The method of taking a real number and rewriting in an easily handled form. One factor must be a number between 1 and 10 and the other an integral power of 10.

EXAMPLES:

1. $12{,}650{,}000 = 1.265 \times 10^7$.
2. $6{,}849{,}000{,}000 = 6.849 \times 10^9$.

Secant

DEFINITION:

In a right triangle, the ratio between the side next to an angle (other than the 90° angle) and the hypotenuse.

EXAMPLE:

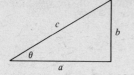

$Secant\ \theta = \dfrac{c}{a}$

Set

DEFINITION:

A collection of elements

EXAMPLES:

1. A is the set [1, 2, 3, 4, 5].

2. B is the set [△, 0, /].

3. C is the set [all girls with blond hair].

Simplify

DEFINITION:

To make easier

EXAMPLE:

$3x^2 + 6x^2 + 9x^2 + x + 3x + 2x$ can be simplified to $18x^2 + 6x$ by adding like terms and $18x^2 + 6x$ can be simplified to $6x(3x + 1)$ by factoring out $6x$.

Sine

DEFINITION:

In a right triangle, the ratio between the side opposite an angle (other than the 90° angle) and the hypotenuse

86 | **Essential Math, Science, and Computer Terms**

EXAMPLE:

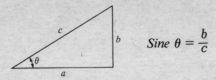

$$\text{Sine } \theta = \frac{b}{c}$$

Note: Cosecant is the reciprocal of the *sine*; cosecant $\theta = \frac{c}{b}$.

Slope

DEFINITION:

The slant of a line or the slant of a line tangent to a circle or arc at a given point. If P_1 is the point (x_1, y_1) and P_2 is the point (x_2, y_2) then the slope of $P_1P_2 = \frac{y_2 - y_1}{x_2 - x_1}$.

EXAMPLES:

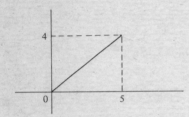

This line has a *slope* of $\frac{y}{x} = \frac{4}{5}$

The *slope of this line is the slope* of the curve at the point *P*.

Solution

DEFINITION:

The answer to a problem

EXAMPLES:

1. $2 + 2 = 4$; 4 is the *solution* to the problem.

2. $\frac{1}{2} + \frac{1}{2} = 1$; 1 is the *solution* to the problem.

Sphere

DEFINITION:

A solid body in space all of whose points are equally distant from a fixed point called the center. (A ball is a sphere.)

EXAMPLE:

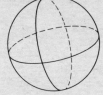

1. Volume of a *sphere* is radius $r = \frac{4}{3}\pi r^3$.

2. The surface area of a *sphere* of radius $r = 4\pi r^2$.

Spiral

DEFINITION:

A curve winding around a point and gradually getting farther away or closer to the point

EXAMPLE:

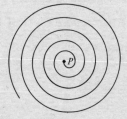

Square

DEFINITION:

1. A number or value multiplied by itself

2. A quadrilateral all of whose sides are equal and whose angles are right angles (geometric)

EXAMPLES:

1. The square of $2 = 2^2 = 2 \times 2$.
   The square of $3 = 3^2 = 3 \times 3$.
   The square of $4 = 4^2 = 4 \times 4$.

2.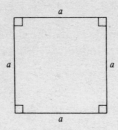

   A geometric square

Square Root

DEFINITION:

Since a *root* is one of the equal factors of an expression, then the *square root* is one of the two equal factors

EXAMPLES:

1. $\sqrt{9}$ The positive *square root* of 9 is 3.
2. $-\sqrt{64}$ The negative *square root* of 64 is -8.
3. $\pm\sqrt{144}$ The two *square roots* of 144 are 12 and -12.
4. $\sqrt{x^2}$ (where $x \geq 0$) = $\pm x$

Statistics

DEFINITION:

A collection of numerical facts or data used to give information about a subject

EXAMPLE:

The following *statistics* are the grades of a class of ten students, used to determine the class average:

Grades

96	86
93	86
91	79
90	70
87	62

Class average is 84.

Substitution (Axiom of Equality)

DEFINITION:

The replacing of one value with another of equal value

EXAMPLES:

1. If $x = y$ and $x + z = d$, then $y + z = d$.
2. If $x = y$ and $xz = d$, then $yz = d$.

Subtend

DEFINITION:

To enclose within certain bounds

EXAMPLE:

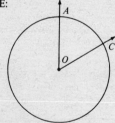

Angle AOC *subtends* the arc AC in this circle.

Subtraction

DEFINITION:

The arithmetic process of finding the difference between numbers. — is the sign used in subtraction

EXAMPLES:

1. $121 - 90 = 31$ *Difference*

2. $\frac{1}{2} - \frac{1}{4} = \frac{1}{4}$ Difference

Sum

DEFINITION:

The result (answer) obtained by adding numbers in an arithmetic problem

EXAMPLES:

1. 44 + 62 = 106; Sum
   The *sum* of 44 plus 62 is 106.

2. 23 + 109 + 61 = 193; Sum
   The *sum* of 23 plus 109 plus 61 is 193.

Supplement

DEFINITION:

When the sum of two angles equals 180°, one angle is said to be the supplement of the other

EXAMPLES:

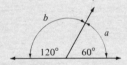

Angle *a* is the *supplement* of angle *b*. (Angle *b* is also the *supplement* of angle *a*.)

Symbol

DEFINITION:

A sign that means something specific

EXAMPLE:

= is the *symbol* meaning equal.
+ is the *symbol* meaning add.
÷ is the *symbol* meaning divide.
π is the *symbol* meaning pi, or 3.14.

Symmetric   (Axiom of Equality)

DEFINITION:

If one value is equal to a second, then the second is equal to the first

EXAMPLES:

1. If $a = b$, then $b = a$.

2. If $(x \cdot y) = (a \cdot b)$, then $(a \cdot b) = (x \cdot y)$.

Symmetry

DEFINITION:

Correspondence of the parts of a figure with reference to a point or line or plane

EXAMPLES:

1. Q (−2, 4) P (2, 4)

P and Q are *symmetric* with respect to the y axis.

In the ellipse: $\dfrac{x^2}{25} + \dfrac{y^2}{100} = 1$

2.
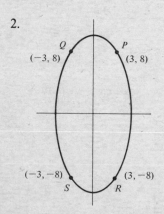

Point P of the ellipse is *symmetric* to:
a. Q wound round the x axis
b. R wound round the x axis
c. S wound round the origin.

Table

DEFINITION:

A listing of values according to certain criteria

Essential Math, Science, and Computer Terms

EXAMPLES:

1. Table of Values of Trigonometric Functions

	sin	cos	tan
1.	.0175	.9998	.0175
2.	.0349	.9994	.0349
3.	.0523	.9986	.0524

2. Table of Squares and Square Roots

N	N^2	$\sqrt{N}$
1	1.0	1.0
2	4.0	1.414
3	9.0	1.732

Tangent

DEFINITION:

1. In a circle or an arc, the line in the plane of the circle that intersects the circle at one and only one point

2. In a right triangle, the ratio between the side opposite an angle (other than the 90° angle) and the side adjacent to that angle

EXAMPLES:

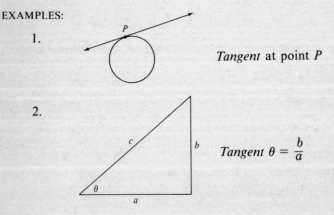

1. *Tangent* at point P

2. *Tangent* $\theta = \dfrac{b}{a}$

Note: Cotangent is the reciprocal of the *tangent*; cotangent $\theta = \dfrac{a}{b}$.

Terms of an Equation

DEFINITION:

The different parts of the equation, which are separated by =, +, or − sign

EXAMPLES:

1. $4x + 3y = 12$
   term   term   term

2. $10x^2 + 6xy + y^2 = 144$
   term   term   term   term

Theorem

DEFINITION:

A statement that can be proved by use of facts and/or definitions, assumptions, and other theorems

EXAMPLES:

1. *Theorem*: The sum of the measures of the angles of a triangle equals 180°.

2. *Theorem*: The shortest segment joining a point to a line is the perpendicular segment.

Transitive   (Axiom of Equality)

DEFINITION:

If one value is equal to a second, and the second value is equal to a third, then the first value is equal to the third.

EXAMPLES:

1. If $x = y$ and $y = z$, then $x = z$.
2. If $(a \cdot b) = (x \cdot y)$ and $(x \cdot y) = (r \cdot s)$ then $(a \cdot b) = (r \cdot s)$.

Translation of Axes

DEFINITION:

    Moving the x and y axes from the normal origin to another point. The new axes are parallel to the old axes.

EXAMPLES:

Normal axes and graph of
$y = -\dfrac{3}{5} x + 3$

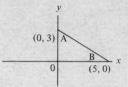

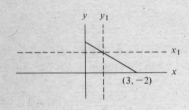

Dotted lines are axes from Fig. 1 translated 2 units to the right and 2 units up (solid axes). Same line now plotted on new translated axes. In terms of new axes, x^1 and y^1, the equation of the line is now $y = -\dfrac{3}{5} x^1 - \dfrac{1}{5}$.

Triangle

DEFINITIONS:

1. A figure in a plane that is the union of three segments determined by three non-colinear points
2. A three-sided polygon

EXAMPLES:

1.

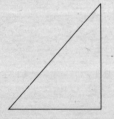

right *triangle*

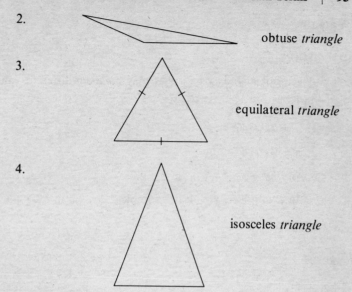

2. obtuse *triangle*

3. equilateral *triangle*

4. isosceles *triangle*

Trigonometric

DEFINITION:

That aspect of mathematics that deals with the sides and angles of triangles and the ways of measuring them

EXAMPLES:

1. $\sin \alpha = \frac{1}{2}$ is a *trigonometric* expression.

2. $\arctan \frac{1}{2} = \theta$ is a *trigonometric* expression.

Value

DEFINITION:

The worth or quantity of an expression

EXAMPLES:

1. The *value* of $10 is equal to the value of 10 one-dollar bills.

2. 6 is the *value* of $5 + 1$.

Variable

DEFINITION:

A symbol such as x or y that may be given different values

EXAMPLES:

1. $x + 2x = 12$
2. $x + 3x = 12$
3. $x + 5x = 12$

In example 1, $x = 4$; in example 2, $x = 3$; in example 3, $x = 2$.

Vector

DEFINITION:

A quantity that has both size and direction

EXAMPLE:

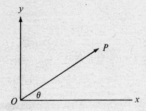

The line OP has a size represented by its length and a direction above the x axis measured by the angle θ.

Vertex

DEFINITIONS:

1. A point in a triangle that is opposite to and the greatest distance from a base of the figure
2. The common end point of two sides of a triangle
3. The point in which the axis of symmetry intersects a graph

EXAMPLES:

1.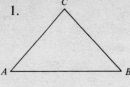

   C is the *vertex* of this triangle in relation to the base line AB.

2.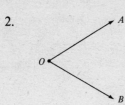

   O is the *vertex* of angle AOB.

3.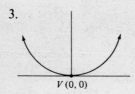

   V (0,0) is the *vertex* of the graph of $y = x^2$.

Vertical

DEFINITION:

At a right angle to the horizontal line

EXAMPLE:

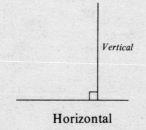

Volume

DEFINITION:

The space enclosed within a three-dimensional figure, measured in cubic units

EXAMPLE:

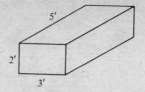

The *volume* of the box is 5 × 2 × 3 = 30 cubic feet.

Math Quiz

1. Match the word in column A with the correct figure in column B.

    Column A   Column B

    1. Angle   (A)

    2. Asymptote   (B)

    3. Cube   (C)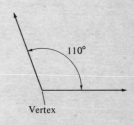

    4. Maximum   (D)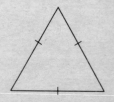

100 | **Essential Math, Science, and Computer Terms**

Column A Column B

5. Minimum   (E)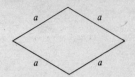

6. Origin   (F)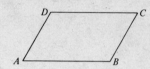

7. Parallelogram (G)

8. Quadrant (H)

9. Radian   (I)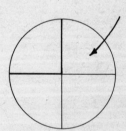

10. Rhombus (J)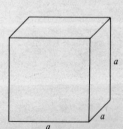

Math Terms | 101

Column A | Column B
11. Triangle | (K)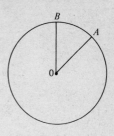

2. Choose the correct word.

(A) Line *a* in the figure is called a(n) (directrix) (ordinate).

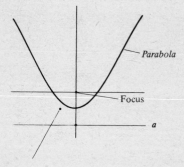

(B) This figure is called a (sphere) (cone).

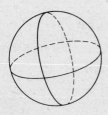

(C) This figure is called a (parabola) (hyperbola).

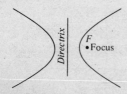

(D) This figure is called a (cone) (chord).

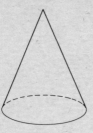

(E) This figure is called a(n) (cube) (ellipse).

(F) This figure is called a (cylinder) (sphere).

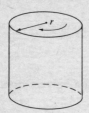

(G) This figure is a (parabola) (hyperbola).

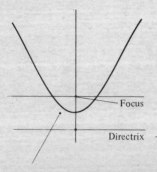

(H) Side c of this right triangle is called a (hypotenuse) (diagonal).

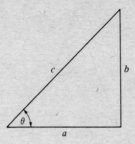

(I) The line in this circle is called a (radius) (diameter).

(J) The line in this circle is called a (radius) (diameter).

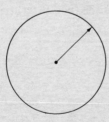

(K) This figure is called a (spiral) (circle).

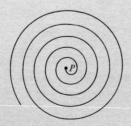

(L) The three points of intersection of the triangle are called (axes) (vertices).

(M) The line in this circle is called a (radius) (chord).

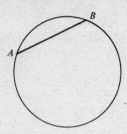

(N) The measure of the boundary ($a + b + c + d + e$) of the figure equals the (perimeter) (circumference).

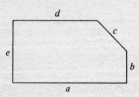

(O) $2\pi r =$ the (perimeter) (circumference).

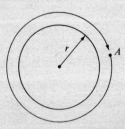

(P) The distance AB is called the (altitude) (ordinate) of this curve. It can also be called the (origin) (amplitude).

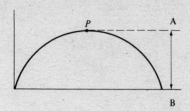

(Q) The → oa is called a (vertex) (vector).

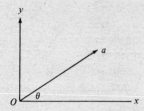

(R) Line a in this figure is called the (diagonal) (chord).

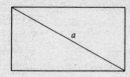

(S) The horizontal distance along the x axis of a graph is called the (ordinate) (abscissa).

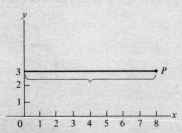

(T) The vertical distance on a graph above the x axis is called the (ordinate) (abscissa).

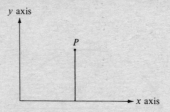

(U) In the right triangle, in relation to angle θ, b/a is called the (tangent) (field).

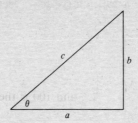

(V) In the right triangle above, in relation to angle θ, $\dfrac{a}{c}$ is called the (sine) (cosine) and $\dfrac{b}{c}$ is called the (cosine) (sine).

3. (A) $3x + x = 12$; is called an _____.

   (B) $a^2 + b^2 = c^2$, when referring to the sides of a triangle, is called the _____ theorem.

   (C) $\sqrt{}$ is called the _____.

4. Give an example of:

   (A) Scientific notation

   (B) Right triangle

(C) The characteristic

(D) The mantissa

(E) An antilogarithm

(F) A dependent variable

(G) An independent variable

(H) A base number

(I) A theorem

(J) A fraction

5. $100 \div 25 = 4$ In this division problem, 4 is the _____ ; 25 is the _____ ; and 100 is the _____ .

6. (A) In the fraction $\frac{48}{6}$, 48 is the _____ and 6 is the _____ .

   (B) $26 \div 7 = 3\frac{5}{7}$. The 5 is called the _____ .

7. (A) $1; -12; \frac{1}{4}; e; \pi; -7.66$ are _____ numbers.

   (B) $4i$ and $4\sqrt{-1}$ are called _____ numbers.

   (C) $|1|; |7|; |-3|$ are _____ values.

   (D) $\int$ is an _____ sign used in calculus.

   (E) $a^6, b^7, c^8, d^{1.6}, e^{-3.1}$. The numerals with each of the above letters are called the _____ of the letter.

8. The square of 3 is _____ . The cube of 7 is
   _____ .

9. (A) The shaded parts of the two figures are called the
   _____ of the figures.

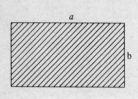
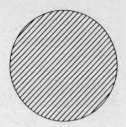

(B) The shaded area of this figure is called the
   _____ of the figure.

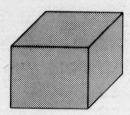

10. The figure in which values may be placed in columns and
rows is called a(n) _____ .

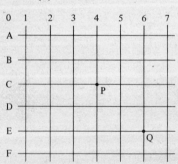

11. The _____ of 9 is 3. The _____ of 27 is 3.

12. (A) 16:40 is called a _____ .

    (B) $\dfrac{dy}{dx} = nx^{n-1}$ is an example of a process called _____ .

    (C) $\dfrac{x}{y} = \dfrac{3}{v}$ is an example of a _____ .

13. Explain the following terms, using either words or pictures:

    (A) Rotate

    (B) Partial fraction

    (C) Revolution

    (D) Scalar

14. Draw a grid system.

15. Give an example of each of the following:

    (A) Logarithm

    (B) Complex number

    (C) Rational number

    (D) Exponent

(E) Polynomial

(F) Function

(G) Coefficient

16. Does *inverse* mean the same as *reciprocal*?

17. What is the difference between:

   (A) *mean* and *median*?

   (B) *oblique* and *obtuse*?

   (C) *normal* and *perpendicular*?

18. What does *symmetry* mean?

19. Give an example of a quadratic equation.

20. Give an example of:

   (A) A set

   (B) A relation

   (C) A function

Math Terms | 111

21. Match the answer from column B with the correct word from column A.

Column A

1. Acute (angle)
2. Associative
3. Circular
4. Congruent
5. Cosecant

6. Coterminal (angles)
7. Decimal point

8. Group
9. Inequality

10. Infinity

Column B

(A) ·

(B) in Fig. 166 (26) c/b is _____ of angle $\propto$

(C) $a + b + c = (a + b + c)$

(D) $2 \neq 3$

(E)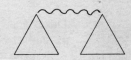

(F) ∞

(G)

(H) $(a)(bc) = (ab)(c)$

(I)

(J)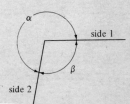

22. Choose the correct word in the following statements:

   1. The arithmetic process of finding the total value of two or more values is called (addition) (subtraction).

   2. if $y = \sin x$, then x is the (cosine) (arc sine) of y.

   3. A decision to accept something as correct without a proof is an (axis) (axiom).

   4. Numbers that have the form $a + bi$ and $a - bi$ are (conjugates) (coordinates) of each other.

   5. The exchanging of a statement and its conclusion is its (converse) (congruent).

   6. A fact that is proved while something else is proved is a (corollary) (coterminal).

   7. The amount by which one value is greater than another is the (determinant) (difference).

   8. A guess about a numerical value is an (estimate) (exponent).

   9. If something is capable of being reached or completed by counting it is (finite) (factorial).

   10. The process of filling in values between other known values is called (interpolation) (identity).

Answers to Math Quiz

Question 1

Question	Answer	Question	Answer
1.	(C)	7.	(F)
2.	(G)	8.	(I)
3.	(J)	9.	(K)
4.	(A)	10.	(E)
5.	(H)	11.	(D)
6.	(B)		

Question 2

Question	Answer	Question	Answer
(A)	directrix	(L)	vertices
(B)	sphere	(M)	chord
(C)	hyperbola	(N)	perimeter
(D)	cone	(O)	circumference
(E)	ellipse	(P)	amplitude, altitude
(F)	cylinder	(Q)	vector
(G)	parabola	(R)	diagonal
(H)	hypotenuse	(S)	abscissa
(I)	diameter	(T)	ordinate
(J)	radius	(U)	tangent
(K)	spiral	(V)	cosine, sine

Question 3

Question	Answer
(A)	equation
(B)	Pythagorean
(C)	square root, radical sign

114 | Essential Math, Science, and Computer Terms

Question 4

Question	Possible Answer
(A)	12,650,000 = 1.265 × 10⁷
(B)	

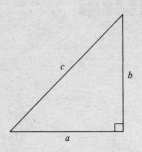

(C)	Log a = 1.6771; 1 is the characteristic of log a.
(D)	Log a = 1.6771; .6771 is the mantissa of log a.
(E)	100 is the antilogarithm of 2 in $\log_{10} 100 = 2$.
(F)	y is the dependent variable in the equation $2x + x = y$.
(G)	x is the independent variable in the equation $2x + x = y$.
(H)	In $10^n = x$, 10 is the base number.
(I)	The sum of the measure of angles in a triangle equals 180°.
(J)	7/12 is a fraction.

Question 5

Question	Answer
(A)	quotient, divisor, dividend

Question 6

Question	Answer
(A)	numerator, denominator
(B)	remainder

Question 7

Question	Answer
(A)	real
(B)	imaginary
(C)	absolute
(D)	integral
(E)	exponents or powers

Question 8

9; 343

Question 9

Question	Answer
(A)	areas
(B)	volume

Question 10

matrix

Question 11

square root; cube root

Question 12

Question	Answer
(A)	ratio
(B)	differentiation
(C)	proportion

Question 13

Question	Answer
(A)	Rotate—to turn or revolve
(B)	Partial—a part of something, as 1/4 is a part of 1/2

116 | Essential Math, Science, and Computer Terms

    (C)   Revolution—the motion of a point or line or curve or surface around a point or axis
    (D)   Scalar—a quantity that can be defined by magnitude alone

Question 14

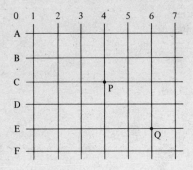

Question 15

Question	Possible Answer
(A)	$\log_{10} 100 = 2$; 2 is the logarithm of 100 (to the base 10)
(B)	$4 + 7i$
(C)	$-1/6, \sqrt{16}, 0, -\sqrt{25}$
(D)	In x^6, 6 is the exponent of x
(E)	$x^2 - 2xy + y^2$
(F)	[(0,0), (1,4), (2,3), (5,8)] is a function
(G)	In $5y$, 5 is the coefficient of y

Question 16

No, multiplicative inverse does.

Question 17

Question	Answer
(A)	*Mean* means the average; *median* means the middle number in a set of numbers ordered according to magnitude.
(B)	*Oblique* means a line that slants *Obtuse* means greater than 90° but less than 180°
(C)	*Normal* and *perpendicular* mean the same.

Math Terms | 117

Question 18

   Symmetry—Correspondence of the parts of a figure with reference to a point or line or plane.

Question 19

$x^2 - 2x - 35 = 0$

Question 20

Question	Answer
(A)	[1, 2, 3, 4, 5] is a set.
(B)	[(0,1,), (2,3), (8,10)] is a relation.
(C)	[(0,1), (1,4), (2,3), (5,8)] is a function.

Question 21

Question	Answer	Question	Answer
1.	(G)	6.	(J)
2.	(H)	7.	(A)
3.	(I)	8.	(C)
4.	(E)	9.	(D)
5.	(B)	10.	(F)

Question 22

Question	Answer	Question	Answer
1.	addition	6.	corollary
2.	arc sine	7.	difference
3.	axiom	8.	estimate
4.	conjugates	9.	finite
5.	converse	10.	interpolation

SCIENCE TERMS

Index of Science Terms

Abrasive
Absolute (Temperature)
Absorb
Acid
Acidic
Adhesion
Affinity
Alkali
Allotropic
Alter
Alternator
Ampere
Analogy
Angstrom
Anode
Anti-
Apparatus
Assumption
Astronomical
Atom
Atomic Number
Atomic Structure
Atomic Weight
Attraction
Audio

Balanced
Barometer

Base
Bond
Buoyancy
By-Product

Calculate
Calorie
Candlepower
Capacitance
Capacitor
Catalyst
Catenary
Cathode
Cavity
Center of Gravity
Centi-
Centrifugal
Charge
Chemical Equation
Circuit
Coaxial
Cohesion
Collision
Combination
Combustion
Component
Composition
Compound

Science Terms | 119

Concave
Concentration
Concept
Condensation
Condenser
Conduct (Electricity)
Conserve
Convection
Converging
Conversion
Convex
Cornea
Corona
Corrosion
Coulomb
Covalent
Critical
Crystalline
Current
Curvature

Damping
Deca-
Decay
Decibel
Declination
Density
Diagram
Dielectric
Diffraction
Diffusion
Dilate
Dilute
Dipole
Displacement
Dissociation
Diverge
Doppler Effect
Dynamic

Elastic
Electrode
Electron
Element
Elementary
Emission
Empirical
Endothermic
End Point
Energy
Environment
Equilibrium
Equivalent
Erg
Evaporate
Exothermic
Expansion
Experiment

Farad
Fatigue
Fermentation
Field
Filament
Filter
Fluid
Flux
Focus
Force
Frequency

Gain
Giga-
Gravity

Harmonic
Henry
Horizon
Horsepower
Hybrid
Hypothesis

Illustrate
Image
Incandescence
Incidence
Inertia
Inhibitor
Insulator
Intensity
Interface
Ion
Isotopes

Kilo-
Kilowatt
Kinetic

Latitude
Level
Longitude
Lumen

Macro
Magnify
Mass
Matter
Medium
Mega-
Melt
Metallic
Micro-
Milli-
Mineral
Mirage
Mixture
Mobility
Mole
Molecule

Nano-
Nuclear

Optical
Oral
Orbit
Oval
Oxidize

Partial
Particle
Period
Periodic
Permeable
Phase (Angle)
Phenomenon
Physical
Pico-
Polarity
Potential
Prediction
Prism
Property
Proton

Qualitative
Quantitative
Quantum

Radial
Radiate
Rate
Ray
Reactance
Reaction
Recession
Rectify
Reference
Reflection
Refraction
Relative
Resonate

Response
Retina
Reversible

Scattering
Schematic
Sideband
Simultaneously
Solution
Solvent
Specific Volume
Spectrum
Spinning
Spontaneous
Standard
State
Substance
Synchronous
Synthetic
System

Technique
Tera-

Terminal
Theory
Thermal
Torque
Transformation
Transition
Transmitting
Transparent
Turbine

Undulation

Vanish
Venturi Effect
Vernier
Vibration
Virtually
Viscosity
Visual
Vocal

Yield

Definitions of Science Terms

Abrasive

DEFINITION:

Something used to grind down something else

EXAMPLE:

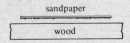

The sandpaper is grinding down the wood. The sandpaper is an *abrasive*.

Absolute   (Temperature)

DEFINITION:

Completely independent of another system or scale

EXAMPLE:

Absolute zero is the lowest temperature possible and equals $-273+°$ C or $-459+°$ F.

Absorb

DEFINITION:

To soak up or swallow up or take in

EXAMPLE:

The sponge *absorbs* the water drops.

Acid

DEFINITIONS:

1. A substance that frees hydrogen ions in solution

2. A substance that contains hydrogen which may be replaced by a metal to form a salt

3. Any compound that can react with a base to form a salt

EXAMPLES:

Examples of *acid* are:

1. In water: $\underset{acid}{H^+ + Cl^-} + \underset{base}{Na^+ + OH^-} \rightarrow H_2O + \underset{salt}{Na^+ + Cl^-}$

2. In benzene: $\underset{base_1}{H_3N:} + \underset{acid}{HCl} \leftrightarrow \underset{base}{[H_3N:H]^+ \ Cl^-}$ (c)

Acidic

DEFINITIONS:

1. Pertaining to a solution or other substance that has all the properties of acid; having a sour taste

2. Any compound that can react with a base to form a salt

EXAMPLES:

1. Silicon is the chief *acidic* element in rocks.

2. *Acidic* action turns litmus paper red.

Adhesion

DEFINITION:

The property of sticking to something else

EXAMPLE:

The paint *adheres* to the wood, but the surface must be clean and dry to get good *adhesion*.

Affinity

DEFINITION:

An attraction between atoms; a force binding atoms

EXAMPLE:

Atom 1   Atom 2

There is an *affinity* between atom 1 and atom 2, and they will combine because of the *affinity*.

Alkali

DEFINITION:

A compound that will react with an acid to give salt plus water

EXAMPLE:

$HCl + NaOH \rightarrow NaCl + H_2O$
acid   *alkali*   salt   water

Allotropic

DEFINITION:

Pertaining to a chemical element that exists in two or more forms which differ in physical properties but have identical chemical formulas

EXAMPLES:

1. The element sulfur can be *allotropic* because as an element in different compounds it can have different physical but identical chemical properties.

2. Carbon in the form of graphite and in the form of diamonds is an *allotropic* element.

Alter

DEFINITION:

To change

EXAMPLE:

Heat *alters* water, changing its form to steam.

Alternator

DEFINITION:

A device that produces alternating currents

EXAMPLES:

1.
2.

Symbols of an *alternator* (AC)

Ampere

DEFINITION:

A unit of electric current equal to 6.281×10^{18} electrons per second. Current is generally shown as I.

EXAMPLE:

$I = \dfrac{E}{R}$, or current is equal to $\dfrac{\text{volts}}{\text{resistance}}$.

Analogy

DEFINITION:

A partial similarity between two things

EXAMPLE:

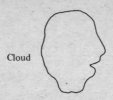

An *analogy* can be made between a cloud and a carefree youth.

Angstrom

DEFINITION:

A very small unit of length equal to one hundred millionth of a centimeter; used mainly in measuring wavelengths of light

EXAMPLE:

Equal to 10^{-10} meter.

Anode

DEFINITION:

The positive electrode

EXAMPLE:

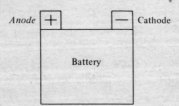

Anti-

DEFINITION:

Against or opposite

EXAMPLES:

1. *Anti*war means against war.

2. *Anti*freeze means preventing freezing.

Apparatus

DEFINITION:

    The equipment or tools necessary to do a job

EXAMPLES:

These objects are some of the *apparatus* a man uses in making a working drawing.

Assumption

DEFINITION:

    A statement accepted as truth without proof

EXAMPLES:

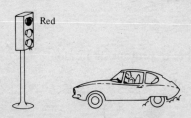

1. I assume (think) the car will stop. I have made an *assumption*.

2. When I see dark rain clouds, I make the *assumption* that it will rain.

Astronomical

DEFINITIONS:

1. Referring to things in the heavens
2. Very high or very large or very great

EXAMPLES:

1. Astronauts are interested in *astronomical* matters.
2. The pressure was *astronomical* (very high).

Atom

DEFINITION:

The smallest part of an element that can be active in a chemical reaction

EXAMPLE:

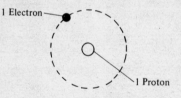

An *atom* of hydrogen

Atomic Number

DEFINITION:

The number of protons in an atom

EXAMPLE:

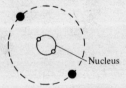

A helium atom has 2 protons in its nucleus; therefore its *atomic number* is 2.

Atomic Structure

DEFINITION:

The two main parts of the atom, i.e., nucleus and electrons, are normally considered to make up the *atomic structure* of an element.

EXAMPLE:

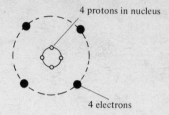

The *atomic structure* of an atom with 4 protons and 4 electrons.

Atomic Weight

DEFINITIONS:

1. The weight of an element, with all its isotopes averaged, as compared with C_{12}
2. The weight of 6.0235×10^{23} atoms of the element

EXAMPLES:

1. The *atomic weight* of hydrogen is 1.00797.
2. The *atomic weight* of oxygen is 15.9994.

Attraction

DEFINITION:

A force on two objects which brings them together

EXAMPLE:

There is an *attraction* between the positive and negative charges.

Audio

DEFINITION:

Having to do with hearing

EXAMPLE:

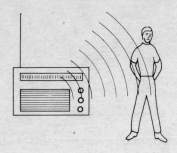

The radio is transmitting *audio* waves, which can be heard by the boy.

Balanced

DEFINITION:

Equal in weight or amount

EXAMPLES:

1. $2 + 2 = 4$ is a *balanced* equation.

2.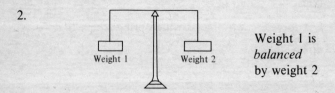

Weight 1 is *balanced* by weight 2

3. $H_2 + Cl_2 = 2HCl$ is a *balanced* chemical equation.

Barometer

DEFINITION:

An instrument used in measuring atmospheric pressure

EXAMPLE:

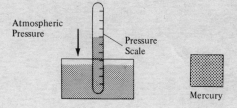

Base

DEFINITION:

Both a substance that forms a salt when it reacts with an acid and one that generally accepts protons

EXAMPLE:

HCl + NaOH = NaCl + H$_2$O

acid   *base*   salt   water

Bond

DEFINITION:

The way in which one atom is joined to another in a chemical compound

EXAMPLE:

```
    H
    |
N — H
    |
    H
```

The chemical *bond* joining the hydrogen atoms to the nitrogen atom is shown by the dash (—).

Buoyancy

DEFINITION:

The upward force exerted upon a body in a liquid

EXAMPLE:

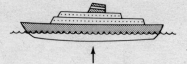

Force of *buoyancy*

By-Product

DEFINITION:

An extra or additional item that comes out of a process designed to form another (more important) item

EXAMPLE:

When coal gas is made, coke is also produced. Coke is a *by-product* of coal-gas manufacture.

Calculate

DEFINITION:

To get an answer, generally by using mathematical means

EXAMPLES:

$764 + 392 + 659 = 1,815$

$62 \times 3 = 186 + 6 = 192 \div 6 = 32$

Examples of *calculations*.

Calorie

DEFINITION:

A basic unit giving the quantity of heat. The amount of

heat required to raise the temperature of 1 gram of water 1 degree centigrade.

EXAMPLES:

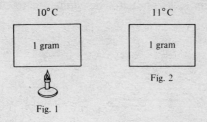

Raising the temperature of the gram of water from 10° C in Fig. 1 to 11° C in Fig. 2 uses 1 *calorie* of heat.

Candlepower

DEFINITION:

The power or intensity that a light has in a certain direction

EXAMPLES:

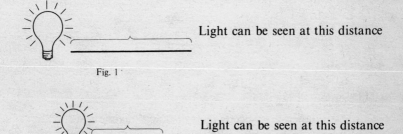

The *candlepower* of the light in Fig. 1 is greater than the *candlepower* of the light in Fig. 2 as it can be seen at a greater distance.

Capacitance

DEFINITION:

The ability of a conductor or condenser to store an electrical charge. Usually shown as C.

EXAMPLE:

Plate A
++++++
Plate B

$C = Q/V =$ charge/potential

Plate A has the *capacitance* to store an electrical charge.

Capacitor

DEFINITION:

An electrical device for storing electricity

EXAMPLE:

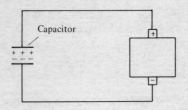

Catalyst

DEFINITION:

A substance that speeds up a chemical reaction but remains unchanged in the process

EXAMPLE:

$$2H_2 + O_2 \xrightarrow{Pt} 2H_2O$$

Platinum (Pt) is a *catalyst* in the reaction between hydrogen (H_2) and oxygen (O_2).

Science Terms | 135

Catenary

DEFINITION:

A curve made in a rope or string or chain when it is not tight and is suspended only at each end

EXAMPLE:

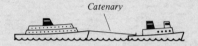

The *catenary* is a chain between two ships.

Cathode

DEFINITION:

The negative electrode

EXAMPLE:

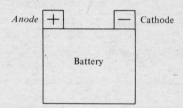

Cavity

DEFINITION:

A hole or hollow place

EXAMPLE:

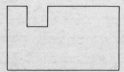

A *cavity* in the box

Center of Gravity

DEFINITION:

The point in a body where the resultant force of weight is located

EXAMPLES:

1. 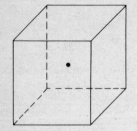 The *center of gravity* is right in the center of the cube.

2. 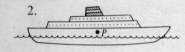 The *center of gravity* of the ship is point *P*, where the total of all resultant forces of weight pass.

Centi-

DEFINITION:

A prefix that means $\frac{1}{100}$ of something (10^{-2})

EXAMPLE:

A *centi*meter is $\frac{1}{100}$ of a meter.

Centrifugal

DEFINITION:

Tending to move away from the center

EXAMPLE:

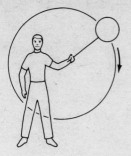

The rotating ball tends to move away from the center and exerts a pulling (*centrifugal*) force on the string

Charge

DEFINITIONS:

1. An atom with an excess of electrons has a negative *charge*. An atom with a lack of electrons has a positive *charge*

2. To replace electrons in a battery (to *charge* it)

EXAMPLES:

1. An oxygen atom lacks an electron, and therefore has a positive *charge*.

2. 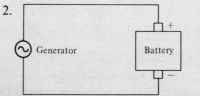 A generator *charging* a battery. It is putting electricity back in the battery.

Chemical Equation

DEFINITION:

A description of a chemical reaction using symbols and elements to show the atoms and molecules in the reaction

EXAMPLE:

$H_2 + Cl_2 = 2HCl$

Circuit

DEFINITION:

A complete path of an electrical current, usually including the generating equipment (battery or generator)

EXAMPLES:

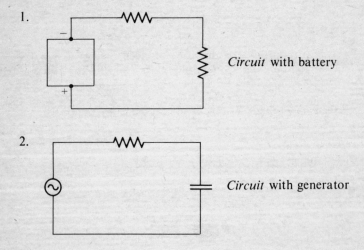

1. *Circuit* with battery

2. *Circuit* with generator

Coaxial

DEFINITION:

Two items, one of which surrounds the other, having the same center

EXAMPLE:

This is a *coaxial* cable since there is an inner cable surrounded by an outer cable.

Cohesion

DEFINITION:

The force holding two things together

EXAMPLE: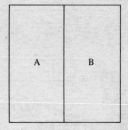

The two blocks are in *cohesion*, held together by glue.

Collision

DEFINITION:

The act of two or more objects hitting each other

EXAMPLE:

The autos just had a *collision*.

Combination

DEFINITION:

The result of joining two or more persons, items, or chemicals

EXAMPLE:

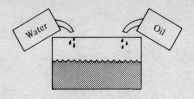

There is a *combination* of water and oil in the tank.

Combustion

DEFINITION:

A burning process

EXAMPLES:

1.  Wood burning is a *combustion* process.

2.  Coal burning is a *combustion* process.

Component

DEFINITION:

A part of something

EXAMPLE:

The *components* of water are hydrogen and oxygen.

Composition

DEFINITION:

An arranging of elements

EXAMPLE:

The *composition* of water is 2 atoms of hydrogen and 1 atom of oxygen, H$_2$O.

Compound

DEFINITIONS:

1. 2 or more elements joined chemically in a substance in definite weight proportions
2. A pure substance made up of more than one element

EXAMPLE:

salt
a *compound* of NaCl

Sodium chloride (NaCl) is a *compound* containing equal numbers of sodium (Na) and chlorine (Cl) atoms.

Concave

DEFINITION:

Hollow and curved inward

EXAMPLE:

Concave

Concentration

DEFINITION:

The amount of a substance located within a certain space or in a quantity of another substance

EXAMPLE:

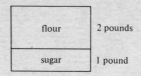

The *concentration* of sugar is $\frac{1}{3}$. The *concentration* of flour is $\frac{2}{3}$.

Concept

DEFINITION:

A thought, an idea, an opinion

EXAMPLE:

This man has a *concept* of a house he is going to build.

Condensation

DEFINITION:

The changing of vapor (steam) into a liquid (water)

EXAMPLE:

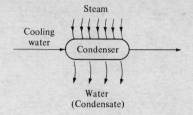

Condenser

DEFINITION:

Equipment used to change a vapor back to liquid

EXAMPLE:

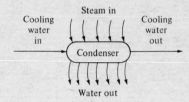

Conduct (Electricity)

DEFINITION:

To lead or guide electricity in a circuit

EXAMPLE:

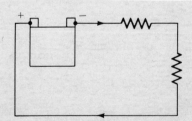

The electricity is being *conducted* from the battery through the circuit and back to the battery.

Conserve

DEFINITION:

To keep as is or to hold down the use of something

EXAMPLE:

Man *conserves* energy by turning off lights.

Convection

DEFINITION:

The transfer of heat through a liquid or gas by the movement of the fluid

EXAMPLE:

The heater warms the air closest to it, which then moves up and away and warms that air.

Converging

DEFINITION:

Coming together toward one point

EXAMPLE:

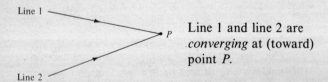

Line 1 and line 2 are *converging* at (toward) point *P*.

Conversion

DEFINITION:

The changing of a substance or a state into another substance or state

EXAMPLE:

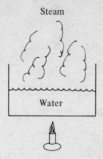

Conversion of water to steam takes place when the temperature is 212° F and the pressure is 14.7 pounds per square inch (p.s.i.).

Convex

DEFINITION:

Curved outwardly and rounded

EXAMPLE:

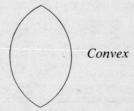

Cornea

DEFINITION:

The part of the eye that admits light and covers the iris and pupil

EXAMPLE:

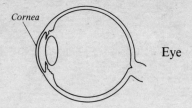

Eye

Corona

DEFINITION:

A bright hazy ring (halo) around the sun

EXAMPLE:

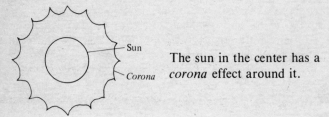

The sun in the center has a *corona* effect around it.

Corrosion

DEFINITION:

Chemical decomposition that takes place on the surface of materials, mainly metals

EXAMPLE:

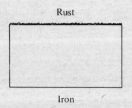

The rust on the iron bar is an example of *corrosion* of iron.

Coulomb

DEFINITION:

The quantity (Q) of electrons transferred by a current (I) of one ampere in one second (t)

EXAMPLE:

$Q = I \times t$

1 coulomb = 1 ampere × 1 second.

Covalent

DEFINITION:

Pertaining to the sharing of electrons by a molecule

EXAMPLE:

$$Ho + Ho \rightarrow H \, {}^{\circ}_{\circ} \, H$$

(hydrogen atoms)   (covalent bond)

In the *covalent* bond, each hydrogen atom contributes one electron.

Critical

DEFINITION:

In general, important or decisive; indicating the place or point at which a change takes place

EXAMPLES:

1. *Critical* mass: the smallest amount of fission material necessary for a nuclear reactor to sustain a chain reaction.

2. *Critical* temperature: the temperature above which a gas cannot be liquefied no matter how much pressure is applied.

3. *Critical* speed: the speed at which unwanted vibrations will enter a machine.

Crystalline

DEFINITION:

A physical state of transparency produced by the orderly, geometrical arrangement of the atoms in a substance

EXAMPLE:

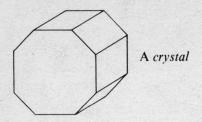

A *crystal*

Current

DEFINITION:

A flow of electrons

EXAMPLES:

1. Electrons appearing to move in a wire.
2. $I \text{ current} = \dfrac{E}{R} \left(\dfrac{\text{Voltage}}{\text{Resistance}} \right)$

Curvature

DEFINITION:

A bending

EXAMPLE:

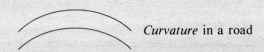

Curvature in a road

Damping

DEFINITION:

Slowing down to reduce the size (amplitude) of a wave motion

EXAMPLE:

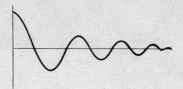

A wave that is being *damped*.

Deca-

DEFINITION:

A prefix that means 10 times something

EXAMPLE:

A *deca*meter is 10 meters.

Decay

DEFINITION:

To change from one substance into another by decomposition

EXAMPLES:

1. A body will *decay* into dust.

2. Wood and leaves may *decay* into coal.

3. Radioactive particles will *decay* into other particles.

Decibel

DEFINITION:

A measure of the intensity (loudness) of sound

EXAMPLE:

$$n = 10 \log_{10} \frac{P_1}{P_2}$$

$n = decibels$; P_1 = intensity of the sound; and P_2 = sound reference level.

Declination

DEFINITION:

The angle from the horizontal to a downward-extending position

EXAMPLE:

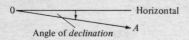

The vector OA has a small angle of *declination*.

Density

DEFINITION:

The mass per unit volume of a substance

EXAMPLE:

The *density* of water is

$$\frac{1 \text{ gram} \quad \text{(weight of mass)}}{1 \text{ cubic centimeter} \quad \text{(unit volume)}}$$

Diagram

DEFINITION:

A drawing used for a scientific, mathematical, or engineering purpose

EXAMPLES:

1.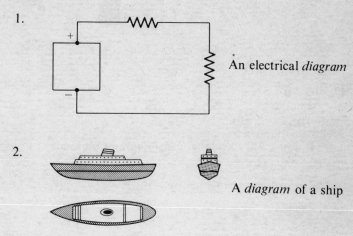

   An electrical *diagram*

2. A *diagram* of a ship

Dielectric

DEFINITION:

A material that does not conduct electricity; an insulator

EXAMPLES:

1. Wood
2. Plastic
3. Mica

Diffraction

DEFINITION:

The bending of a ray of light as it strikes the edge of an

object; a change that light goes through as it goes by the edges of opaque bodies or narrow slits

EXAMPLE:

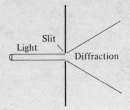

Diffusion

DEFINITIONS:

1. For gases: the equal distribution of molecules within the walls of a vessel

2. For light: the scattering of light

EXAMPLES:

1.

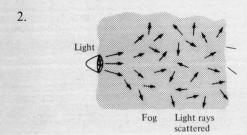

2.

Dilate

DEFINITION:

To make larger or to increase in volume; to expand

Science Terms | 153

EXAMPLE:

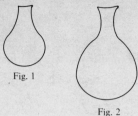

Fig 2 shows the bag of Fig. 1 *dilated* to twice its size.

Dilute

DEFINITION:

To make less concentrated. Generally done by adding water to a substance

EXAMPLE:

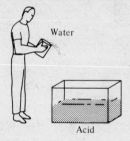

To *dilute* the acid, the man is pouring water into the acid.

Dipole

DEFINITION:

A molecule having equal numbers of positive and negative charges, with the charges arranged in such a way that one part of the molecule is predominantly positive and the other is predominantly negative

EXAMPLE:

H • 0 •   Water molecules form *dipoles*.

+   H

Displacement

DEFINITIONS:

1. The removing of a substance from its usual place
2. The amount (volume) of water moved by a floating body
3. The difference between one place location and another

EXAMPLES:

1.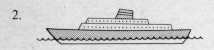

   The pump is *displacing* the water from the tank.

2.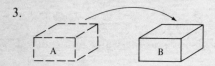

   The ship *displaces* the amount of water below the water level.

3.

   The box was *displaced* from position A to position B.

Dissociation

DEFINITION:

The breakup, usually reversible, of a molecule into atoms or similar molecules

EXAMPLE:

$H_2O = H^+ + OH^-$

The *dissociation* of water results in hydronium (H^+) and hydroxyl (OH^-).

Diverge

DEFINITION:

To bend or separate in different directions

EXAMPLE:

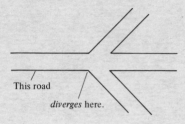

Doppler Effect

DEFINITION:

A change in sound (frequency) as a noise gets closer or farther away

EXAMPLE:

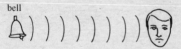

As the bell gets closer to the ear, the frequency changes (the *Doppler Effect*).

Dynamic

DEFINITION:

Moving; energy; forceful changes in motion. The opposite of *dynamic* is static.

EXAMPLE:

This moving car is in *dynamic* motion.

Elastic

DEFINITION:

Capable of being stretched and then springing back

EXAMPLE:

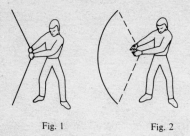

Fig. 1   Fig. 2

The man pulls (stretches) the rubber in Fig. 1. It springs back in Fig. 2 because it is *elastic*.

Electrode

DEFINITION:

A conductor for passing electric current in or out of a cell or body

EXAMPLE:

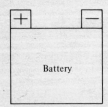

Electrodes. The positive *electrode* is the anode. The negative *electrode* is the cathode.

Electron

DEFINITION:

A negative (−) particle

EXAMPLE:

Electricity is a flow of *electrons*.

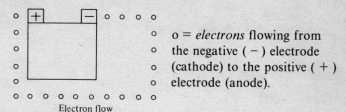

o = *electrons* flowing from the negative (−) electrode (cathode) to the positive (+) electrode (anode).

Electron flow

Element

DEFINITION:

A substance that is entirely made up of atoms that have the same atomic number

EXAMPLES:

Element	Atomic Number
1. Aluminum	13
2. Boron	5
3. Carbon	6
4. Fluorine	9
5. Gold	79
6. Iodine	53
7. Iron	26
8. Mercury	80
9. Oxygen	8
10. Silver	47

Elementary

DEFINITION:

Basic, simple

EXAMPLE:

$1 + 1 = 2$ is an *elementary* math problem.

Emission

DEFINITION:

Energy sent out or discharged

EXAMPLES:

1. 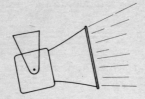 An *emission* of sound from a loudspeaker.

2. 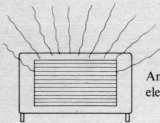 An *emission* of heat from an electric heater.

3.  An *emission* of light from a bulb.

Empirical

DEFINITION:

Answers or results based on experiments or observations alone, not theory

EXAMPLE:

We know that bees can fly because we see them do so. This is an *empirical* fact.

Endothermic

DEFINITION:

Pertaining to action in which heat is absorbed

EXAMPLES:

1. Water evaporating from my skin makes me feel cold. This is an *endothermic* reaction.

2. When an egg boils, an *endothermic* action takes place; the egg yolk absorbs heat, which makes it harder.

End Point

DEFINITION:

The point at which there is a change in the physical or chemical property of a solution

EXAMPLES:

1. 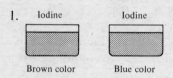 When the *end point* is reached in mixing starch solution in iodine, the solution's color changes from brown to blue.

2. 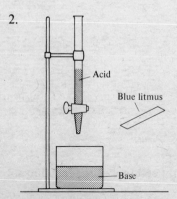 At the *end point* when the amount of acid added equals the amount of base, the litmus paper turns red.

Energy

DEFINITION:

The ability to do work

EXAMPLES:

1. Potential energy  Weight may fall

2. Kinetic energy  Weight moving through air

3. Electrical energy  Electrons moving and turning a light on

4. Heat energy 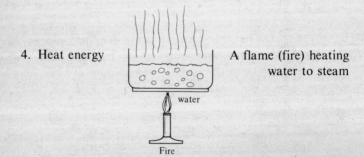 A flame (fire) heating water to steam

Science Terms | 161

Environment

DEFINITION:

The external conditions and material surroundings

EXAMPLE:

The *environment* includes the air, water and land. The smoke from the house is hurting the *environment*.

Equilibrium

DEFINITION:

A state of equality between forces

EXAMPLE:

The two boys are in a state of *equilibrium*.

Equivalent

DEFINITION:

The same as

EXAMPLE:

2 is *equivalent* to 1 + 1.

Erg

DEFINITION:

The unit of work or energy in the centimeter, gram,

second system; the work done by a force of 1 dyne acting through a distance of 1 centimeter

EXAMPLE:

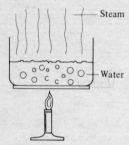

Evaporate

DEFINITION:

To change from a liquid into a vapor (gas)

EXAMPLE:

Water *evaporates* into steam.

Exothermic

DEFINITION:

Pertaining to a process in which heat is given off or made (generated)

EXAMPLES:

1. The burning of a candle is an *exothermic* reaction.

2.  In the atomic blast, there is an *exothermic* action; heat is generated along with the explosion.

Expansion

DEFINITION:

The process of making larger

EXAMPLE:

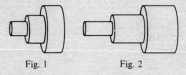

Fig. 1   Fig. 2

Fig. 2 is the result of an *expansion* of Fig. 1.

Experiment

DEFINITION:

A trial to prove or disprove something

EXAMPLE:

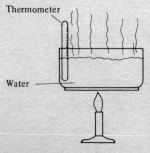 In this *experiment*, water is heated to see if it will turn to steam at 100° C.

Farad

DEFINITION:

The unit of electrical capacity *(f)*

EXAMPLE:

If a condenser is charged to a potential (E) of 1 volt by 1 coulomb (Q) of electricity, it has an electrical capacity or capacitance (C) of *farad*.

$$C = \frac{Q}{V} \qquad 1 \; farad = \frac{1 \; coulomb}{1 \; volt}$$

Fatigue

DEFINITION:

The state of having lost energy or power

EXAMPLE:

The iron bar is showing *fatigue*, which is shown by the cracks in it.

Fermentation

DEFINITION:

A chemical change in organic matter caused by the action of an enzyme

EXAMPLE:

Yeast in flour (Fig. 1) makes it rise when baked (Fig. 2). The yeast causes *fermentation* in the flour.

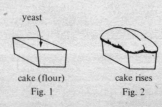

cake (flour)　　　cake rises
Fig. 1　　　　　　Fig. 2

Field

DEFINITIONS:

1. A region or area created by electricity or magnetism wherein a force exists
2. A gravitational *field* is a region in which a body having mass exerts a force of attraction on another large body

EXAMPLES:

1.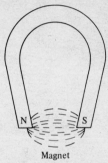

    A *field* (dotted line) exists because of the magnetic force between the N and S poles.

2.  The moon is in the earth's gravitational *field*.

Filament

DEFINITION:

A wire (usually in an electronic tube) that when heated gives off electrons or heats a cathode that gives off electrons

EXAMPLE:

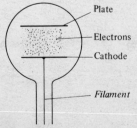

Filter

DEFINITION:

1. A device used to purify liquids or gases by removing suspended matter
2. A device used to absorb certain wave lengths of light and allow others to pass

EXAMPLES:

1.

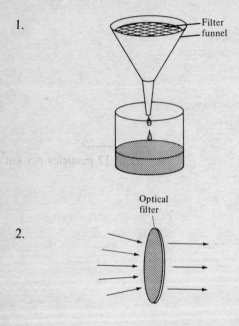

2.

Fluid

DEFINITION:

A substance that is not solid and can flow; a substance that takes the shape of the vessel containing it

EXAMPLES:

Water is a *fluid*.
Gases are *fluids*.
Oil is a *fluid*.

Science Terms | 167

Flux

DEFINITIONS:

1. A substance that causes other substances to flow
2. A material used to help in soldering
3. The number of particles in a flowing steam or beam
4. The number of parts in a unit volume

EXAMPLES:

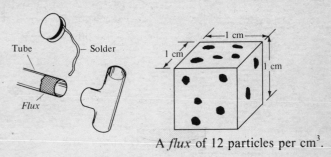

A *flux* of 12 particles per cm^3.

Focus

DEFINITION:

The point at which rays meet

EXAMPLE:

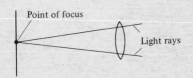

Force

DEFINITION:

A push or pressure that acts on an object and changes the state of rest or motion

168 | Essential Math, Science, and Computer Terms

EXAMPLES:

1. 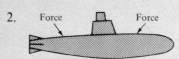 The man is exerting *force* on the mower.

2.  The weight of the water is exerting a *force* on the submarine.

Frequency

DEFINITION:

The number of cycles, periods, or vibrations that is repeated, usually within a given period of time

EXAMPLES:

1.

The *frequency* of the spikes is 8.

2.

The *frequency* of a sine wave.

Gain

DEFINITION:

To increase or make something get larger

EXAMPLES:

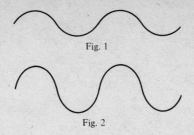

Fig. 1

Fig. 2

The wave in Fig. 2 has *gained* in strength, from its size in Fig. 1.

Giga-

DEFINITION:

A word which means 10^{+9}

EXAMPLE:

100 *giga*cycles equals $100 \times 10^{+9} = 10^{11}$ cycles.

Gravity

DEFINITION:

The force of attraction exerted on an object by the earth

EXAMPLE:

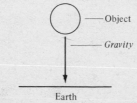

Force pulling object to earth.

Harmonic

DEFINITIONS:

1. Motion along a line around a center point
2. A multiple of a number

EXAMPLES:

1.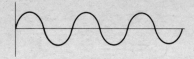

Harmonic motion

2. 300 is the third *harmonic* of 100; 400 is the fourth *harmonic* of 100.

Henry

DEFINITION:

The unit of inductance (L)

EXAMPLE:

If a back electromotive force of 1 volt is induced in a coil by a current that varies at a rate of 1 amp per second, the inductance is 1 *henry*.

Horizon

DEFINITION:

The line where earth or ocean and sky meet

EXAMPLE:

Sky
────────── *Horizon*
Ocean

Horsepower

DEFINITION:

Power equal to 550 ft.-lbs./sec. Also, 1 *horsepower* = 745.7 watts.

EXAMPLE:

10 *Horsepower* = 5,500 ft.-lbs./sec.
     OR
10 *Horsepower* = 7,457 watts

Hybrid

DEFINITION:

> The result in the breeding of different races, breeds, varieties or species

EXAMPLE:

> A mule is a *hybrid* of an ass and mare (female horse).

Hypothesis

DEFINITION:

> A thought put forth because of observed facts; a theory that has not been fully proved

EXAMPLE:

> That there is life on Mars is a popular *hypothesis*.

Illustrate

DEFINITION:

> To make something clear or understood by showing or doing

EXAMPLE:

> The man is *illustrating* how to duck for his son.

Image

DEFINITION:

A form showing a real person or object

EXAMPLE:

Incandescence

DEFINITION:

The emission (sending) out of light caused by high temperatures

EXAMPLES:

1. An *incandescent* light bulb.

2. A fluorescent light sending out its *incandescence*.

Incidence

DEFINITION:

The ray of light or projectate represented by a vector that falls on a surface

EXAMPLE:

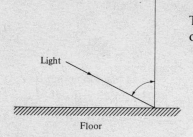

The angle of *incidence* of the ray of light falling on the floor.

Inertia

DEFINITION:

The tendency of a thing or person to stay in a state of rest or uniform motion in the same direction unless acted upon by some external force

EXAMPLE:

The car will continue to go forward of its own *inertia* unless a force or brake stops it.

Inhibitor

DEFINITION:

A substance that slows down a chemical reation

EXAMPLE:

Oil is an *inhibitor* of rust.

Insulator

DEFINITION:

A material that prevents heat or electricity from passing

EXAMPLES:

1. 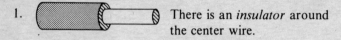 There is an *insulator* around the center wire.

2. 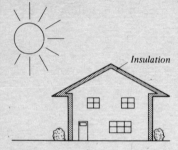 The *insulation* helps keep the heat out of the house.

Intensity

DEFINITION:

The amount of strength

EXAMPLE:

Fig. 1

Fig. 2

The *intensity* of the light in Fig. 2 is greater than the *intensity* of the light in Fig. 1.

Interface

DEFINITION:

A common boundary between two or more surfaces

EXAMPLES:

1. The solid area is the *interface* between the two boxes.

2.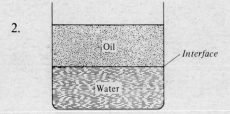

Ion

DEFINITION:

An electrically charged atom, radical, or molecule having an excess or deficiency of electrons

EXAMPLES:

1. Na^+ = an *ion* of an atom.
2. SO_4^{-2} = a radical *ion*.
3. H_3O^+ = an *ion* of a molecule
4.

    An oxygen atom with only one electron is a positive *ion*.   An oxygen atom with 3 electrons is a negative *ion*.

Isotopes

DEFINITION:

Atoms of an element having the same number of protons in the nucleus or the same atomic number, but having a different number of neutrons in the nucleus or a different atomic weight

EXAMPLES:

1. $^{17}_{8}O$ $^{11}_{7}N$ Two oxygen *isotopes* with different weights.

2. $^{18}_{8}O$ $^{12}_{7}N$ Two nitrogen *isotopes* with different numbers of neutrons in their nucleus.

Kilo-

DEFINITION:

A prefix that means 1,000 times something (10^3)

EXAMPLE:

A *kilo*watt is 1,000 watts.

Kilowatt

DEFINITION:

1,000 (kilo) watts. A watt is one volt × one amp. A *kilowatt* is equal to 1,000 volts × one amp.

EXAMPLE:

3 *kilowatts* = 3 kw

Science Terms | 177

Kinetic

DEFINITION:

Related to motion

EXAMPLES:

1. 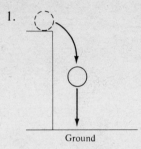 A falling object has *kinetic* (motion) energy.

2.  This moving car has *kinetic* (motion) energy.

Latitude

DEFINITION:

Angular distance from a circle or plane of reference

EXAMPLE:

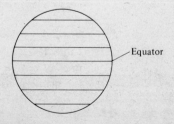

The lines of *latitude* are parallel to the equator.

Level

DEFINITION:

Even with or horizontal

EXAMPLES:

1.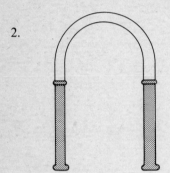

The two boards are *level*.

2.

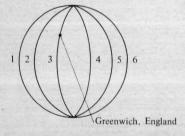

The two columns are *level*.

Longitude

DEFINITION:

Angular distance from the prime meridian, which goes through Greenwich, England. Lengthwise.

EXAMPLE:

Lines 1, 2, 4, 5, and 6 are lines of *longitude* parallel to the prime meridian.

Science Terms | 179

Lumen

DEFINITION:

A unit of light; the amount of light emitted in one second by a light source of one candle

EXAMPLE:

Light

$1 = d$
1 second of time

Plate

The amount of light falling on the plate in 1 second after traveling a distance of 1 unit.

Macro

DEFINITION:

Very large

EXAMPLE:

A *macro* (very large) view of a tack.

Magnify

DEFINITION:

To enlarge or make bigger

EXAMPLE:

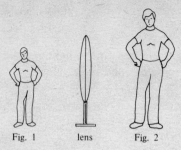

Fig. 2 is a *magnified* image of the man in Fig. 1.

Mass

DEFINITIONS:

1. A quantity of matter

2. The quantity of matter in a body measured by comparing the changes in speeds (velocity) that result when the body and a standard body strike each other

EXAMPLES:

1. Dirt

The *mass* or volume of dirt on the ground.

2. $F = Ma$

The M in this familiar equation is mass.

Matter

DEFINITION:

What any physical object is composed of; it occupies space

Science Terms | 181

EXAMPLES:

1.

The sand is composed of small particles that take up space and are therefore *matter*.

2.

Man is composed of *matter*; he is a physical object that occupies space.

Medium

DEFINITION:

An intervening substance through which a force acts or an effect is produced

EXAMPLE:

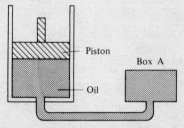

Oil is the *medium* through which the pressure from the piston is transmitted to box A.

Mega-

DEFINITION:

A prefix that means one million times something (10^6 42)

EXAMPLE:

*Mega*ohm means 1 million ohms.

Melt

DEFINITION:

To change from a solid to a liquid state

EXAMPLES:

1.

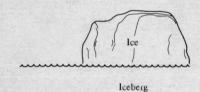

The heat of the sun *melts* the iceberg.

2.

The fire *melts* the wax of the candle.

Metallic

DEFINITION:

Made of metal

EXAMPLES:

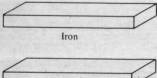

1. The iron bar is *metallic*.

2. The steel bar is *metallic*.

Micro-

DEFINITION:

1. Very small
2. One millionth part of a unit

EXAMPLE:

A *micro*ohm is $1/10^6$ ohm.

Milli-

DEFINITION:

A prefix that means $\frac{1}{1,000}$ of something (10^{-3})

EXAMPLE:

A *milli*amp is $\frac{1}{1,000}$ of an ampere.

Mineral

DEFINITION:

An inorganic chemical element or compound that occurs naturally in the earth

EXAMPLES:

1. Calcium sulfate
2. Iron oxide
3. Sodium chloride
4. Gold
5. Platinum

Mirage

DEFINITION:

A vision that is imagined

EXAMPLE:

A thirsty man on a desert sometimes thinks he sees water when it does not exist. What he sees is a *mirage*.

Mixture

DEFINITION:

Two or more substances that are put together but do not combine chemically. The substances retain their individual properties

EXAMPLES:

1.  These boys and girls are *mixed* together.

2. 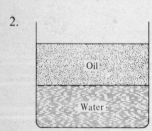 The container holds a mixture of oil and water. The oil and water, although in the same container, do not mix chemically and retain their own properties.

Mobility

DEFINITION:

The ability to move about

EXAMPLE:

An ambulance has great *mobility*, as it can move about easily.

Mole

DEFINITIONS:

1. Mass equal to its molecular weight in grams
2. A unit of substance that contains as many units as there are atoms in 12 one gram of carbon 12

EXAMPLE:

One *mole* of H_2O has a mass of 18.02 g and contains 6.02×10^{23} molecules.

Molecule

DEFINITIONS:

1. The chemical union of two or more like or unlike atoms
2. The smallest unit of matter that retains chemical identity in mass
3. The smallest part of a substance that moves around as a whole

EXAMPLES:

1. H_2O A *molecule* of water.
2. O_2 A *molecule* of oxygen.

Nano-

DEFINITION:

A prefix that means 10^{-9}

EXAMPLE:

A frequency of 1 million *nano*cycles is equal to 1,000.0 $\frac{1}{10^{-9}} = \frac{1}{10^{-3}}$ cycles.

Nuclear

DEFINITION:

Having to do with reactions involving the nucleus (center) of an atom

EXAMPLE:

Atom

The center of the atom is its nucleus, which is made up of protons and neutrons. When there is an internal reaction, it is called a *nuclear* reaction.

Optical

DEFINITION:

Having to do with the transmission, absorption, reflection, or refraction of light

EXAMPLES:

1.

2.

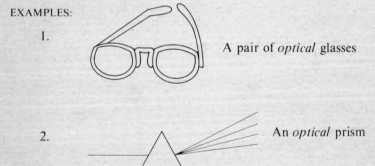

A pair of *optical* glasses

An *optical* prism

Oral

DEFINITIONS:

1. Spoken

2. Having to do with the mouth
3. Taken into the mouth

EXAMPLES:

1. The teacher gave an *oral* presentation (by talking) of his research.
2. The child took his medicine *orally* (by mouth).

Orbit

DEFINITION:

The path taken by one body as it revolves around another

EXAMPLES:

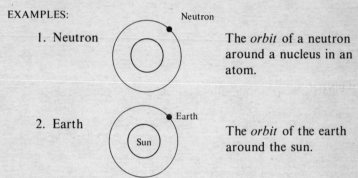

1. Neutron — The *orbit* of a neutron around a nucleus in an atom.

2. Earth — The *orbit* of the earth around the sun.

Oval

DEFINITION:

Having the form of an egg

EXAMPLE:

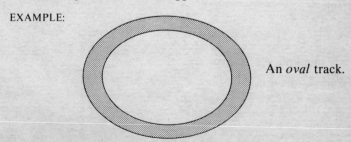

An *oval* track.

Oxidize

DEFINITIONS:

1. To have a substance react with oxygen
2. To join something with oxygen
3. To take away hydrogen from a substance

EXAMPLES:

1. $2Fe + 3O_2 \rightarrow 2Fe_2O_3$ The reaction shows how iron is *oxidized*.
2. $N + O_2 \rightarrow NO_2$ The nitrogen has been joined by oxygen (*oxidized*) to give nitrogen oxide.

Partial

DEFINITION:

Part of something; not complete

EXAMPLE:

The picture is only a *partial* work (not completely done).

Particle

DEFINITION:

A small piece of something

EXAMPLES:

1. A *particle* of dirt.
2. A *particle* of cloth.

Period

DEFINITION:

> The amount of time required for completion of a cycle or for the cycle to begin to repeat itself

EXAMPLE:

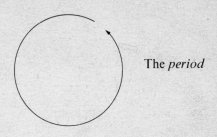

The *period*

Periodic

DEFINITION:

> Occurring at regular intervals

EXAMPLE:

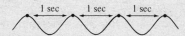

> The peaks (highest points) of the wave are *periodic*, occurring exactly one second apart.

Permeable

DEFINITION:

> Permitting substances to pass through; porous

EXAMPLE:

> A filter is *permeable*. Water can *permeate* (go through) the filter.

Phase (Angle)

DEFINITION:

The angle between two vectors that represent two different changing quantities (such as current and voltage) with the same frequency

EXAMPLE:

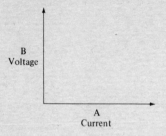

Vector B, representing voltage, is 90° out of phase with vector A, representing current. The phase angle is 90°.

Phenomenon

DEFINITION:

A scientific fact or event that has a description or explanation

EXAMPLES:

1. A comet in the sky is a *phenomenon*.
2. A nuclear reaction is a *phenomenon*.

Physical

DEFINITION:

Pertaining to the material characteristic of something

EXAMPLE:

The size, weight, and color of a substance are some of its *physical* traits.

Pico-

DEFINITION:

A prefix that means 10^{-12}

EXAMPLE:

A frequency of 1 million *pico*cycles is equal to
$1,000,000 \times \dfrac{1}{10^{-12}} = \dfrac{1}{10^{-6}}$

Polarity

DEFINITION:

The charge of a magnetic pole

EXAMPLES:

+ is a positive *polarity*.

− is a negative *polarity*.

Potential

DEFINITIONS:

1. Having the possibility of doing something
2. The difference in force (electric V or E or EMF)

EXAMPLES:

1. The rock has the *potential* of falling.

2.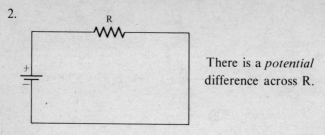

There is a *potential* difference across R.

Prediction

DEFINITION:

A statement that something is going to happen in the future

EXAMPLES:

1. He made a *prediction* that it would rain tomorrow.

2. The statesman made a *prediction* that war would end next year.

Prism

DEFINITION:

A solid figure whose bases or ends are similar, equal, and parallel polygons and whose faces are parallelograms

EXAMPLE:

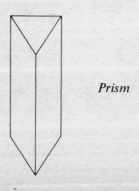

Prism

Property

DEFINITION:

A feature or trait or characteristic of something

EXAMPLES:

1. A *property* of steel is that it is hard.

2. A *property* of wood is that it will burn.

3. A *property* of water is that it is wet.

Proton

DEFINITION:

A particle that is found in the nucleus of all atoms and comprises the atomic nucleus of the protium isotope of hydrogen. A proton carries a unit positive charge equal to the negative charge of an electron. It has a mass of 1.672×10^{-24} gram.

EXAMPLE:

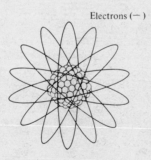

Electrons (−)

Nucleus (composed of neutrons and *protons* (+))

Qualitative

DEFINITION:

Having to do with the nature of something and not its numerical value

EXAMPLES:

1. Balls are made of good, hard rubber. (This is a *qualitative* statement about balls.)
2. *Qualitative* analysis of the ore shows that it contains iron, silicon, and sulfur.

Quantitative

DEFINITION:

Having to do with actual numbers or value

EXAMPLE:

The *quantitative* value of the balls in the can is 12.

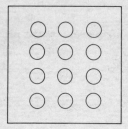

Quantum

DEFINITION:

The unit amount of energy released or absorbed during the emission or absorption of radiations

EXAMPLE:

Quantum theory: energy exists in discrete units, only whole numbers of which can exist; each unit is called a *quantum*.

Radial

DEFINITION:

Having parts like rays coming from a center

EXAMPLE:

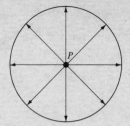

Radial lines of force coming from point *P*.

Radiate

DEFINITION:

To send out rays

EXAMPLE:

The fire *radiates* heat.

Rate

DEFINITION:

The amount of something changed per unit of time

EXAMPLES:

1. The *rate* of a bill for electricity is measured in kilowatts of electricity per hour.

2. The *rate* of speed of a racing car might be 115 miles per hour.

Ray

DEFINITIONS:

1. A straight line of energy traveling from its source to a given point

196 | Essential Math, Science, and Computer Terms

2. A beam of light that comes from a bright source

EXAMPLES:

1.

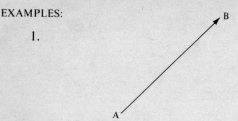

Line *l* is a ray going from point *A* to point *B*.

2.

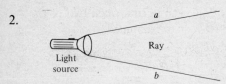

The ray is represented by the beam contained within lines *a* and *b*.

Reactance

DEFINITION:

A property of alternating current that is due to capacitance *(C)*, or inductance *(L)*, or both. It is expressed in ohms and labeled X.

EXAMPLE:

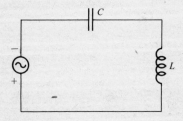

The *reactance* is composed of value of L and C in a special formula.

$X_L = 2\pi f L$ where X_L = *reactance* in ohms

$X_C = \dfrac{1}{1\pi f C}$ X_C = *reactance* in ohms $\pi = 3.146$

Note:

$\dfrac{1}{f}$ = frequency in cycles.

L = inductance in henrys.
C = inductance in farads.

Reaction

DEFINITIONS:

1. A chemical change that occurs when two or more different substances are combined

2. The force that tends to oppose an applied force

EXAMPLE:

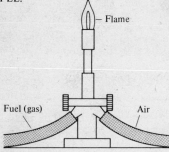

The *reaction* of the fuel to air produces a flame.

Recession

DEFINITION:

The act of going away or back from something

EXAMPLES:

1. The man is *receding* from the house.

2.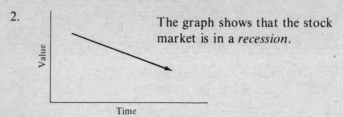

The graph shows that the stock market is in a *recession*.

Rectify

DEFINITION:

To make something correct, right, or straight

EXAMPLE:

Fig. 1   Fig. 2

The picture in Fig. 1 was hung crookedly. The mistake was *rectified* in Fig. 2; the picture now hangs straight.

Reference

DEFINITIONS:

1. A place (point) from which to start
2. A source (book or article) of information

EXAMPLES:

1.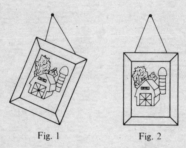

The *reference* point is point P.

2. The *Handbook of Chemistry* is a *reference* book.

Reflection

DEFINITION:

The return of light or rays from a surface

EXAMPLE:

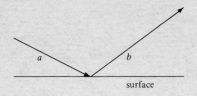

Ray *a* is a *reflection* of ray *b* off the surface.

Refraction

DEFINITION:

A light ray's change of direction in passing obliquely from one medium to another

EXAMPLE:

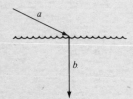

Refraction changes light ray *a* to light ray *b* as it goes from air to water.

Relative

DEFINITION:

As compared with

EXAMPLES:

1.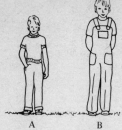

Boy A is *relatively* smaller than boy B.

2. 1.00001 is *relatively* larger than 1.00000.

Resonate

DEFINITION:

To vibrate, in response to sound waves, at the same frequency as or at a multiple frequency of the sound waves

EXAMPLE:

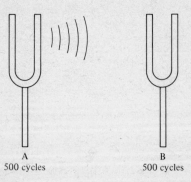

Tuning fork A, when struck, will cause tuning fork B to *resonate* since the two forks have the same frequency (500 H).

Response

DEFINITION:

An answer

EXAMPLE:

Why? Because I'm sick.

Man B is giving a *response* to man A.

Retina

DEFINITION:

The part of the eye that receives the image and is connected to the brain

EXAMPLE:

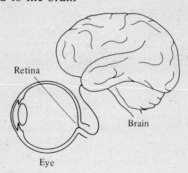

Reversible

DEFINITION:

Capable of returning to its first (original) state

EXAMPLE:

$H_2O_l \leftrightarrow H_2O_g$

Water changing to steam and back again is a *reversible* action.

Scattering

DEFINITION:

Spreading out

EXAMPLE:

Fig. 1   Fig. 2

In Fig. 1, the balls are together. Fig. 2 shows the balls after *scattering*.

Schematic

DEFINITION:

A simple drawing or sketch of something showing its function or successive actions

EXAMPLE:

A *schematic* diagram of an electrical circuit.

Sideband

DEFINITION:

The frequencies that lie on either side of a radio (carrier) wave having an audio component part

EXAMPLE:

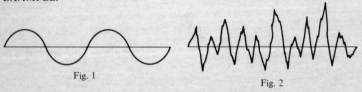

Fig. 1   Fig. 2

Fig. 1 represents an unmodulated carrier wave. Fig 2 shows the carrier wave with an audio input on it. The frequencies above and below the unmodulated carrier wave are the *sideband*.

Science Terms | 203

Simultaneously

DEFINITION:

Done at the same time

EXAMPLE:

All the men are throwing a ball *simultaneously*.

Solution

DEFINITION:

A mixture of two or more dissimilar (different) substances that results in one substance having one similar (same) set of properties

EXAMPLE:

Adding salt (NaCl) to water (H_2O) produces a *solution* of salt water.

$$H_2O + SO_3 \rightarrow H_2SO_4$$

Solvent

DEFINITION:

A substance that is capable of dissolving other substances in it

EXAMPLE:

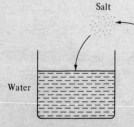

When salt is put into water, the water dissolves the salt. The water is the *solvent*.

Specific Volume

DEFINITION:

The volume occupied by one gram of a substance at a particular temperature and pressure. It is the reciprocal of density.

EXAMPLE:

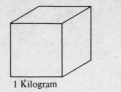

1 Kilogram

$$SV = \frac{cm^3}{Kg}$$

Spectrum

DEFINITION:

The different components formed when light or electromagnetic rays are broken up into different colors or frequencies

EXAMPLE:

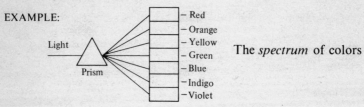

The *spectrum* of colors

Spinning

DEFINITION:

Rotating or turning around an axis

EXAMPLES:

1. The earth is *spinning* on its axis.
2. The earth is *spinning* around the sun.

Spontaneous

DEFINITION:

Happening by itself, without outside help

EXAMPLE:

A pile of hot oily rags bursts into flame without a spark or match being placed by it. This is a *spontaneous* fire.

Standard

DEFINITION:

A recognized form indicating quality or quantity; the main or basic rule of measurement of a particular action or system

EXAMPLES:

1. The *standard* atmospheric pressure is 14.7 p.s.i.
2. The *standard* temperature is 0° C.

State

DEFINITION:

The physical condition of a substance or matter

EXAMPLE:

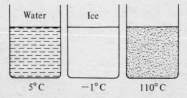

The *states* of H_2O: liquid, solid, gas.

Substance

DEFINITION:

A particular kind of matter

EXAMPLE:

The *substances* in bread include water and flour.

Synchronous

DEFINITIONS:

1. Happening at the same
2. In phase with each other

EXAMPLE:

The two waves are *synchronous* (in phase).

Synthetic

DEFINITION:

Something not real or genuine

EXAMPLE:

The tires were made of *synthetic* rubber. The substance does not come from a rubber plant but from chemicals mixed in such a way as to appear to be rubber.

System

DEFINITION:

A grouping of objects or items united or joined by some interaction or dependence on each other to perform some function

EXAMPLE:

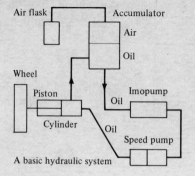

A basic hydraulic *system*.

Technique

DEFINITION:

A way of doing something

EXAMPLES:

1. A boxer may have a certain *technique* in throwing his fists.

2. A painter may have a certain *technique* in applying paint.

Tera-

DEFINITION:

A prefix that means 10^{12}

EXAMPLE:

1,000 *tera*cycles equal $1,000 \times 10^{+12} = 10^{+15}$ cycles.

Terminal

DEFINITIONS:

1. The end or boundary

2. A connection on a battery

EXAMPLES:

1.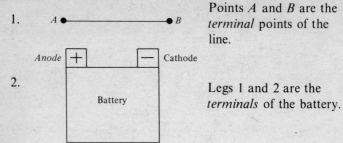

Points *A* and *B* are the *terminal* points of the line.

2.

Legs 1 and 2 are the *terminals* of the battery.

Theory

DEFINITION:

A thought or idea, not proved, that explains something

EXAMPLE:

Men first thought that the earth was flat; later, some had a *theory* that it was round. The *theory* that the earth was round was proved correct.

Thermal

DEFINITION:

Hot; having to do with heat

EXAMPLE:

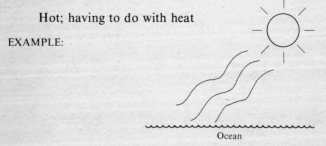

The sun is having a *thermal* (heating) effect on the ocean.

Torque

DEFINITION:

A force that makes something rotate (turn)

EXAMPLE:

The man is putting a *torque* on the shaft by turning it with a wrench.

Transformation

DEFINITION:

The process of changing something into something else

EXAMPLE:

Heat causes the *transformation* of water into steam.

Transition

DEFINITION:

A change from one place (or thing) to another

EXAMPLES:

1. The *transition* from feet to metric scale: 3.39 feet = 1 meter.

2. There is a *transition* from liquid to gas when water is heated to 100° C at 14.7 pounds p.s.i.

Transmitting

DEFINITION:

The sending of something

EXAMPLES:

1.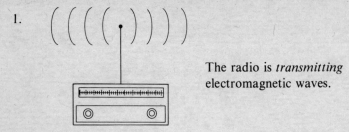

The radio is *transmitting* electromagnetic waves.

2. The automobile's design calls for *transmitting* power from the engine to the rear wheels.

Transparent

DEFINITION:

Capable of being seen through

EXAMPLE:

A glass window is *transparent*; therefore you can see through it.

Turbine

DEFINITION:

A machine in which a shaft is turned when air, steam, or water hits a ring or rings of blades attached to the shaft

EXAMPLE:

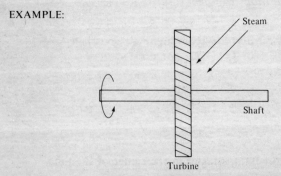

Undulation

DEFINITION:

The process of rising up and down, as in a wave

EXAMPLE:

The *undulation* of the wave is shown above.

Vanish

DEFINITION:

To go away, to disappear

EXAMPLE:

Fig. 1　　　　Fig. 2

The man on the right side in Fig. 1 has *vanished* in Fig. 2.

Venturi Effect

DEFINITION:

A speeding-up effect that occurs when a fluid is sent through a tube that is narrower in the middle than at the ends

EXAMPLE:

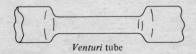

Venturi tube

The speed of fluid is faster in the center of the tube.

Vernier

DEFINITION:

A scale that measures with a very high degree of accuracy. This is usually done by having a marked *(vernier)* scale slide alongside a scale that has larger divisions.

EXAMPLE:

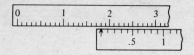

The *vernier*-scale arrow shows that the answer is 1.8 plus some more. By looking at where the lines on the *vernier*-scale most closely line up with lines on the larger scale, you can see that the .5 is just in line with a larger scale line. The answer is therefore 1.85.

Vibration

DEFINITION:

Rapid motion in alternating directions

EXAMPLES:

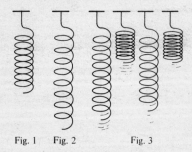

Fig. 1   Fig. 2   Fig. 3

In Fig. 1, the spring is at rest. In Fig. 2, it is pulled. In Fig. 3, it is let go and starts its *vibration* up and down.

Virtually

DEFINITION:

Almost the same but not exactly

EXAMPLE:

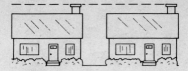

The two buildings are *virtually* the same size.

Viscosity

DEFINITION:

The ease or difficulty with which a fluid flows

EXAMPLE:

Oil flows more slowly than water. Therefore, it has greater *viscosity*.

Visual

DEFINITION:

Having to do with sight

EXAMPLES:

1. A *visual* discovery is a discovery made by eye.

2. A *visual* inspection is done by examining something by eye.

Vocal

DEFINITIONS:

1. Having to do with the voice

2. Making a sound with one's voice

EXAMPLE:

The crowd was very vocal (made a loud sound with their voices).

Yield

DEFINITION:

To give way

EXAMPLE:

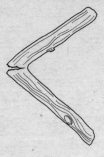

The stick is at its *yield* point, the point at which it is cracking (giving way).

Science Quiz

1. Match the word in column A with the correct figure in column B.

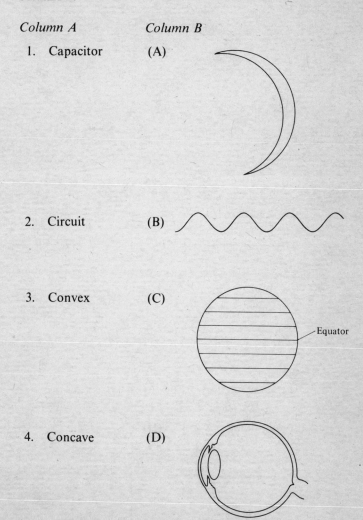

   Column A

   1. Capacitor
   2. Circuit
   3. Convex
   4. Concave

216 | **Essential Math, Science, and Computer Terms**

Column A	Column B
5. Cornea	(E) 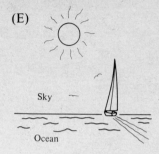
6. Cavity	(F) 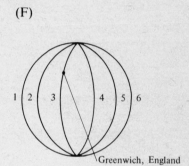
7. Latitudinal	(G) 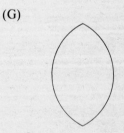
8. Longitudinal	(H) 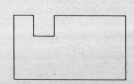

Science Terms | 217

Column A   Column B
 9. Horizon   (I)
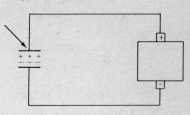

10. Undulation   (J)
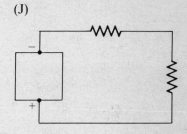

2. Match the definition in column A with the correct word in column B.

   Column A   Column B

   (A) To swallow up completely   1. affinity
       or take in

   (B) An attraction between atoms   2. anode

   (C) The way one atom is joined   3. atomic
       to another   number

   (D) To change something   4. absorb

   (E) The positive electrode   5. barometer

   (F) The smallest part of an element that is   6. conduct
       in a chemical reaction

   (G) To lead or guide electricity   7. ampere
       in a circuit

218 | Essential Math, Science, and Computer Terms

Column A   Column B

(H) Instrument used in measuring   8. alter
    atmospheric pressure

(I) Unit of measure of   9. bond
    electric current

(J) The number of protons in an atom   10. atom

3. Choose the correct word.

   (A) The changing of a vapor into a liquid is called (evaporation) (condensation).

   (B) The changing of one substance into another is called (conversion) (conduction).

   (C) Arrangement of elements is called (composition) (critical).

   (D) Change of one substance to another is called (density) (decay).

   (E) Dynamic means (moving) (stopped).

   (F) A negative particle is a(n) (electron) (proton).

   (G) The rate at which something is sent out is (energy) (emission).

   (H) A push or pressure that acts on an object or changes the state of rest or motion is (work) (force).

   (I) The number of times something is repeated is its (frequency) (amplitude).

   (J) A material that prevents heat or electricity from escaping is (insulation) (expression).

Science Terms | 219

4. There are four kinds of energy or ability to do the work pictured below. Match the type of energy with a picture representing that energy:

(A) Potential energy   1.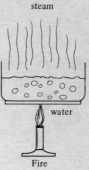

(B) Kinetic energy   2.

(C) Electric energy   3.

(D) Heat energy   4.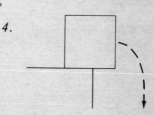

5. Choose the correct word.

   (A) An electrically charged atom that has fewer electrons than necessary is an (ion) (image).

   (B) To change from a solid to a liquid is to (melt) (oxidize) a substance.

(C) Two or more substances put together is a (concentration) (combination).

(D) A piece of equipment used to change steam back to water is a (circuit) (condenser).

(E) The path of something around something else is its (orbit) (period).

(F) The amount of time required to finish or complete a cycle is its (amplitude) (period).

(G) The features or traits of something are its (elements) (properties).

(H) An ampere per second is a (coulomb) (henry).

(I) A bending downward is a (declination) (elevation).

(J) When you make something larger you (expand) (yield) it.

6. Choose the word that best answers the question.

(A) The two lines in the figure below are (diverging) (converging).

(B) The rounded part of this figure is its (oval) (curvature).

(C) This figure is a (crystal) (electron).

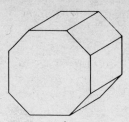

(D) The three roads in this figure (diverge) (exchange).

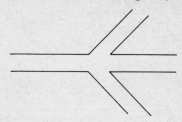

(E) The point at which the rays in the figure meet is its (doppler) (focal) point.

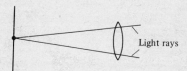

(F) The figure in the mirror is the (image) (incidence) of the man.

(G) The bold area in the picture is the (interface) (filter) of the two surfaces.

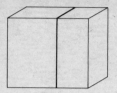

(H) The rocks in Fig. 1 have been (conserved) (scattered) in Fig. 2.

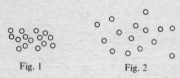

(I) When something has the form of an egg it is (oval) (conical) in shape.

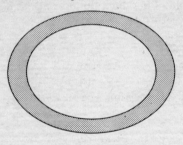

(J) This figure is a (prism) (anode).

7. Choose the correct word in each example:

(A) This is an example of being (radial) (oval).

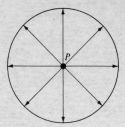

(B) Ray *a* is a (reflection) (refraction) of ray *b*.

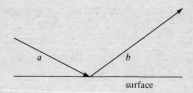

(C) Ray *b* is a (refraction) (reflection) of ray *a*.

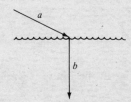

(D) The antenna in this figure below is (terminating) (transmitting) waves.

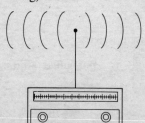

(E) The two weights in the picture are (balanced) (bonded).

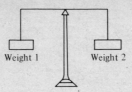

(F) The two cars have had a (compound) (collision).

(G) The two men on the board are (balanced) (based).

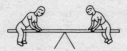

(H) The force in this figure is (centrifugal) (centrifical).

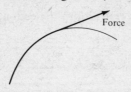

(I) The steel bar in the figure is (fatigued) (forced).

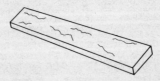

8. Answer true (T) or false (F).

  (A) *A substance* is a particular kind of matter.
     (T) (F)

  (B) A substance that slows up a chemical reaction but remains unchanged is a *catalyst*. (T) (F)

(C) An *inhibitor* is a substance that speeds up a chemical reaction.   (T)   (F)

(D) A substance that forms a salt when it reacts with an acid and generally accepts protons is a *base*.
   (T)   (F)

(E) A difference between two things is an *analogy*.
   (T)   (F)

(F) A solution that has basic properties of salt is *acidic*.
   (T)   (F)

(G) The negative electrode is the *cathode*.   (T)   (F)

(H) When electrons share a molecule, they are *covalent*.
   (T)   (F)

(I) A *critical* item is not an important item.   (T)   (F)

(J) Changing one substance to another is a *conversion*.
   (T)   (F)

9. Choose the correct word in each example:

   (A) (Rate) (Density)   is the mass per unit volume.

   (B) (Displacement) (Bond)   is the amount (volume) of water removed by a floating body.

   (C) (Energy) (Flux)   is the ability to do work.

   (D) A(n)   (mole) (erg)   is the unit of energy in the gram, centimeter, second system.

   (E) (Photons) (Isotopes)   are atoms of the same element having the same atomic number but a different mass and a different number of neutrons in the nucleus.

(F) An (ion) (anode) is an electrically charged atom.

(G) (Macro) (Matter) is what any physical object is composed of and occupies space.

(H) (Mass) (Macro) is a quantity of matter.

(I) (Metallic) (Melt) means made of metal.

(J) (Mineral) (Solvent) is a chemical element that occurs normally in the earth.

(K) The (ion) (center of gravity) is the center point of something.

(L) A(n) (mole) (atom) is a mass equal to its molecular weight in grams.

(M) A (farad) (photon) is electromagnetic radiation generated when a particle having an electric charge changes its momentum in collision with electrons.

(N) To (alter) (oxidize) is to join something with oxygen.

(O) A (current) (bond) is a flow of electrons.

10. Match the words in column A with the picture or definition in column B.

Column A   Column B

(A) Circuit diagram   1.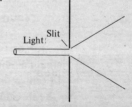

Science Terms | 227

Column A *Column B*

(B) Dielectric   2.

(C) Diffraction   3. A measure of capacitance

(D) Diffusion   4.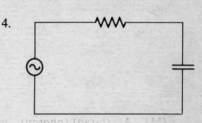

(E) Doppler Effect   5.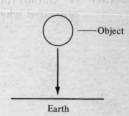

(F) Farad   6. 550 ft.-lbs./sec.

(G) Experiment   7. A unit of energy released

(H) Gravity   8. Wood, mica, plastic

(I) Horsepower   9.

                              Container

(J) Quantum   10. A trial to prove or disprove something

Answers to Science Quiz

Question 1

Question	Answer	Question	Answer
1.	(I)	6.	(H)
2.	(J)	7.	(C)
3.	(G)	8.	(F)
4.	(A)	9.	(E)
5.	(D)	10.	(B)

Question 2

Question	Answer	Question	Answer
(A)	4	(F)	10
(B)	1	(G)	6
(C)	9	(H)	5
(D)	8	(I)	7
(E)	2	(J)	3

Question 3

Question	Answer	Question	Answer
(A)	condensation	(F)	electron
(B)	conversion	(G)	emission
(C)	composition	(H)	force
(D)	decay	(I)	frequency
(E)	moving	(J)	insulator

Question 4

Question	Answer	Question	Answer
(A)	4	(C)	2
(B)	3	(D)	1

Question 5

Question	Answer	Question	Answer
(A)	ion	(F)	period
(B)	melt	(G)	properties
(C)	combination	(H)	coulomb
(D)	condenser	(I)	declination
(E)	orbit	(J)	expand

Question 6

Question	Answer	Question	Answer
(A)	converging	(F)	image
(B)	curvature	(G)	interface
(C)	crystal	(H)	scattered
(D)	diverge	(I)	oval
(E)	focal	(J)	prism

Question 7

Question	Answer	Question	Answer
(A)	radial	(F)	collision
(B)	reflection	(G)	balanced
(C)	refraction	(H)	centrifugal
(D)	transmitting	(I)	fatigued
(E)	balanced		

Question 8

Question	Answer	Question	Answer
(A)	(T)	(F)	(F)
(B)	(F)	(G)	(T)
(C)	(F)	(H)	(T)
(D)	(T)	(I)	(F)
(E)	(F)	(J)	(T)

Question 9

Question	Answer	Question	Answer
(A)	density	(I)	metallic
(B)	displacement	(J)	mineral
(C)	energy	(K)	center of gravity
(D)	erg	(L)	mole
(E)	isotopes	(M)	photon
(F)	ion	(N)	oxidize
(G)	matter	(O)	current
(H)	mass		

Question 10

Question	Answer	Question	Answer
(A)	4	(F)	3
(B)	8	(G)	10
(C)	1	(H)	5
(D)	9	(I)	6
(E)	2	(J)	7

COMPUTER TERMS
Index of Computer Terms

Accumulator
Address
Algorithm
Alphanumeric
Alternative
Analog
Approximate
Array
Assembler (Language)
Automatic

Basic
Batch
Binary
Bit
Block
Block Diagram
Boolean
Buffer
Byte

Capacity
Cassette
Central Processing Unit (CPU)
Character
Classify
Cobol

Coding
Collate
Compiler
Computer Language
Console
Control
Core

Debugging
Device
Digital (Computer)
Dimension
Display

Edit
Error
Execute
Explicit
Expression (Computer)

Feedback
File
Finite
Fixed Point
Floating Point
Flow Chart
Fortran

Generic

Hardware
Hexidecimal

Index
Infinite
Input
Integer
Interpreter
Inventory
Invert
Iterate

Keypunch

Logic
Loop
Lower Case

Magnetic Disk
Magnetic Drum
Magnetic Tape
Matrix (Computer)
Memory
Method
Mode

Output

Precedence
Program
Punch Card

Random Access
Real Time (Program)
Record
Registers
Remote Job Entry
Repetitive
Retrieval
Routine

Scan
Schedule
Sequence
Sequential Device
Software
Sort
Specification
Statement
Stochastic
Storage (Store)
Structure
Subroutine

Terminal
Time Sharing
Tolerance
Transaction
Transform
Translator
Truncate

Upper Case

Variable

Definitions of Computer Terms

Accumulator

DEFINITION:

A part of a computer used to add things together

EXAMPLE:

X = X + Y(I)

Fortran statement showing X as the *accumulator* for the Y(I).

Address

DEFINITION:

A number used to represent a location, especially for stored data

EXAMPLE:

2, 4 could be used as the *address* for this point.

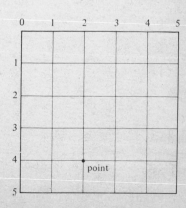

Algorithm

DEFINITION:

A step-by-step simplified way (procedure) for getting a certain answer (result)

EXAMPLE:

Algorithm for boiling an egg:

1. Get egg.
2. Get pot of water.
3. Put egg in water.
4. Turn on gas.
5. Wait for water to boil 3 minutes.
6. Remove egg.

Alphanumeric (Data)

DEFINITION:

Data consisting of letters, symbols, and/or numbers, but in a form such that calculations cannot be performed on them

EXAMPLES:

1. YES
2. R2D2
3. BLACK

Alternative

DEFINITION:

A different choice

EXAMPLE:

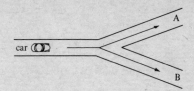

The car has the *alternative* of going along road A or going along road B.

Analog

DEFINITION:

A type of computer that uses physical quantities (voltages, distances, etc.) to represent more general measurements

EXAMPLE:

A slide rule is a very simple *analog* computer.

Approximate

DEFINITION:

Close to something

EXAMPLES:

1. 1.1 is *approximately* equal to 1.1001.

2.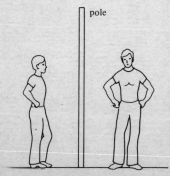

The men are in the *approximate* area of the pole.

Array

DEFINITION:

A set of items (especially data) arranged in a meaningful way

EXAMPLES:

1.

A	B	C	D
1	2	3	4

2.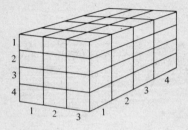

Assembler (Language)

DEFINITION:

A programming language that requires one instruction for every machine-language instruction. (Also the name of the program.)

EXAMPLES:

Assembler Instruction	*General Meaning*
L 3, 0(3)	Load data to register 3.
SR 3, 7	Subtract data in register 7 from data in register 3.
AR 2, 3	Add data in register 3 to data in register 2.

Automatic

DEFINITION:

Happening by itself without (direct) outside help

EXAMPLE:

When the elevator is full, the closing of the doors is *automatic*.

Basic

DEFINITION:

A computer programming language used mainly for time sharing and mini-computer applications

EXAMPLE:

Basic statements:

```
READ A, B, C
LET X = (A−B)/C
PRINT X
DATA 1234, 489, 8
END
```

Batch

DEFINITIONS:

1. To put things together.
2. To execute a series of computer programs one after another.

EXAMPLE:

Fig. 1   Fig. 2

The separated items in Fig. 1 were *batched* together in Fig. 2.

Binary

DEFINITION:

A numbering system that is based on 2's instead of 10's and uses only the digits 0 and 1

EXAMPLES:

Decimal Number		Binary Number
0	=	0
1	=	1
2	=	10
3	=	11
4	=	100

Bit

DEFINITION:

A way of giving one of two choices, generally "yes" or "no," "go" or "not go," "on" or "off"

EXAMPLE:

The number 1 or 0 can be used in a computer to mean "yes" or "no."

Block

DEFINITIONS:

1. A group of things put together

2. A rectangle or a rectangular solid

EXAMPLES:

1.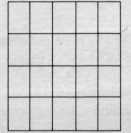

A *block* of stamps.

2. 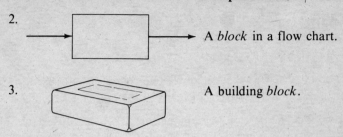 A *block* in a flow chart.

3. 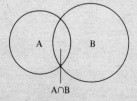 A building *block*.

Block Diagram

DEFINITION:

A picture that shows in a simple way the general arrangement of a system (piece of equipment)

EXAMPLE:

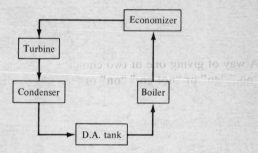

Block diagram of a simple steam system.

Boolean

DEFINITION:

A special kind of algebra used in the study of sets

EXAMPLE:

$A \cap B$ *(Boolean* notation for the intersection of sets A and B).

240 | **Essential Math, Science, and Computer Terms**

Buffer

DEFINITIONS:

1. An area that separates two things, usually used to slow or stop movement between the two
2. The part of a computer core through which data must pass when it is read from or written to external devices

EXAMPLE:

The wall acts as a *buffer* between the two men.

Byte

DEFINITION:

The smallest addressable unit of data storage in a computer, usually made up of 8 bits

EXAMPLE:

See *Bit*.

Capacity

DEFINITION:

The amount that something can hold or take

EXAMPLES:

1.

The *capacity* of the can is one gallon.

2. The *capacity* of a computer is the amount of storage space contained in the memory.

Cassette

DEFINITION:

A role of tape enclosed within a case

EXAMPLES:

1.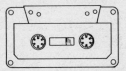

A *cassette* of tape used in a tape recorder.

2.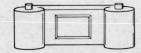

A *cassette* of film used in a camera.

Central Processing Unit (CPU)

DEFINITION:

The main part of a computer

EXAMPLE:

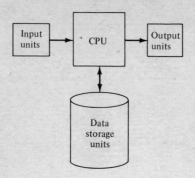

Character

DEFINITION:

1. A letter, digit, or symbol
2. Any item produced by depressing a typewritten or keypunch key

EXAMPLE:

R2-D2 consists of five characters.

Classify

DEFINITION:

To put an item into a group (class) with other similar items; this is often done by assigning a code

EXAMPLES:

1. All boys names will be *classified* by a *B* in front of the name.
2. All athletes will be *classified* by the number 1 after their names.

Cobol

DEFINITION:

A computer programming language used mostly for business applications

EXAMPLES:

>Common
>Business
>Oriented
>Language

Coding

DEFINITION:

>A method of representing data

EXAMPLE:

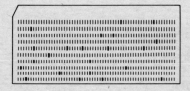

>The holes in this punch card are *coding* and represent data.

Collate

DEFINITION:

>To put things together in the right order, especially the pages of a book or report

EXAMPLES:

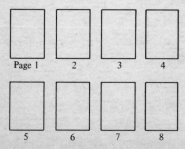

You *collate* each page in order to get a complete set.

Complete set of all eight pages

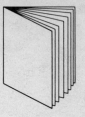

Compiler

DEFINITION:

    A special program that changes statements from one language into another, usually into machine language

EXAMPLES:

    1. Fortran *compiler*

    2. Cobol *compiler*

    3. Basic *compiler*

    4. PL-1 *compiler*

Computer Language

DEFINITION:

    The combination of words and phrases through which a person can give instructions to a computer

EXAMPLES:

    Names of different languages:

        Fortran
        Cobol
        PL-1
        Assembler

Console

DEFINITION:

    The part of a computer where keys and other devices that control the computer are located

EXAMPLE:

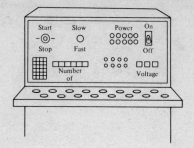

Control

DEFINITION:

To regulate or be in charge of

EXAMPLE:

Policemen *control* traffic.

Core

DEFINITIONS:

1. The center or middle of anything
2. The individual, small magnetic rings that are used for data storage in some computers
3. The part of a computer used to store data and program instructions at the time the program is executed

EXAMPLE:

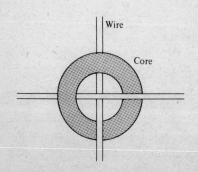

Debugging

DEFINITION:

Finding and correcting program errors

EXAMPLE:

Fortran Statement:

ERROR (BUG): $X = 4*(A(I) + B(J)$
DEBUGGED FORM: $X = 4*(A(I) + B(J))$

Device

DEFINITION:

A piece of equipment that does a specific job

EXAMPLE:

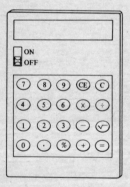

The calculator is a *device* used to do mathematical work.

Digital (Computer)

DEFINITION:

A computer that uses data in the form of numbers rather than physical values

EXAMPLES:

1. IBM 360 computers.

2. Univac 1110 computers.

3. CDC 6800 computers.

Dimension

DEFINITION:

One part of a measurement

EXAMPLE:

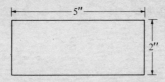

5" is the long *dimension*.
2" is the short *dimension*.

Display

DEFINITION:

To show something visually. This may take the form of tables, graphs, drawings, etc.

EXAMPLE:

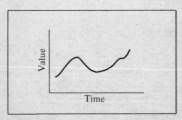

You can *display* a value-vs.-time graph on a screen (cathode ray tube).

Edit

DEFINITION:

To examine data to see if it is correct

248 | Essential Math, Science, and Computer Terms

EXAMPLE:

The man is *editing* a role of tape by looking at it through a magnifying glass.

Error

DEFINITION:

A mistake; something that is not right or correct

EXAMPLE:

$1 + 1 = 3$ is an *error*.

Execute

DEFINITIONS:

1. To carry out (instructions)
2. To cause (programmed instructions) to be carried out

EXAMPLE:

The man is starting the computer so he can *execute* the program.

Explicit

DEFINITION:

Exact, with no doubt left as to its meaning

EXAMPLE:

A very *explicit* statement from a mother to a child:

"I want you to go to your room right now, remove everything from your pockets, undress, take a shower, put your pajamas on, and go directly to bed."

Expression   (Computer)

DEFINITION:

Constants, variables, and functions that are connected by symbols and punctuation to describe a computation

EXAMPLE:

(3A + C) / (B*B) * (cos θ)

Feedback

DEFINITION:

The putting back of something into a system that has gone through the system as a way of altering future output of that system

EXAMPLE:

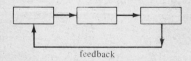

File

DEFINITION:

An ordered collection of similar information recor

EXAMPLE:

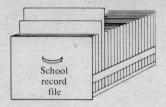

File of student information in a school, with one record for each student

Finite

DEFINITION:

Having limits or an end or a last number

EXAMPLES:

1.

There are a *finite* number of apples in this box.

2. The English alphabet has a *finite* number (26) of letters.

Fixed Point

DEFINITION:

A way of storing numbers without storing decimal points; the decimal point is known to be located in a fixed position

EXAMPLES:

Number	Fixed Point Form
123	123000
2.897	2897
456.7	456700

↑
Known Decimal Position

Floating Point

DEFINITION:

A way of representing data (especially on computer cards) so that the location of the decimal point is shown separately for each item of data

EXAMPLES:

Floating Point	Fixed Point
1.2345	12345
832.	8320000
9845.	98450000

Flow Chart

DEFINITION:

A "picture" of a program, showing the things to be done and the order in which they are to be done

EXAMPLE:

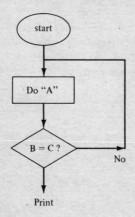

Fortran

DEFINITION:

*For*mula *Tran*slator, a computer programming language used mainly for scientific applications

EXAMPLES:

Fortran Statements:

```
READ (5,10) A,B,C
X = (A-B)/C
WRITE (6,11) X
STOP
END
```

Generic

DEFINITION:

Pertaining to the next (generally improved) kind of an item or piece of equipment

EXAMPLE:

> Fortran I
> Fortran II
> Fortran III
> Fortran IV

An example of a *generic* improvement in computer programming. We started with Fortran I and constantly improved it until now we have Fortran IV.

Hardware

DEFINITIONS:

1. Physical tools and equipment, especially those made of metal
2. The actual equipment used in and with computers

EXAMPLES:

1. Hammers, nails, etc.
2. The computer, keypunch machines, disks, input/output machines.

Hexadecimal

DEFINITION:

A number system using 16 as the base. The digits used are: 0, 1, 2, 3, 4, 5, 6, 7, 8, 9, A, B, C, D, E, and F (A = 10, B = 11, C = 12, D = 13, E = 14, F = 15).

EXAMPLES:

Decimal	Hexadecimal
15	F
24	18 (1 × 16 + 8)
44	2C (2 × 16 + 12)

Index

DEFINITION:

A listing (usually alphabetical) that shows where subjects are located

EXAMPLE:

Item	*Page*
Algorithm	1
Amplitude	18
Analytical	7
Combinations	36
Directrix	110

Infinite

DEFINITION:

Having no limits or end or last number

EXAMPLE:

There are an *infinite* number of numbers (no end); you can keep on counting forever.

Input

DEFINITION:

The data or information to be taken from an external (outside) source and put into the computer

EXAMPLE:

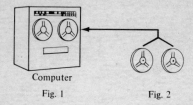

Computer
Fig. 1   Fig. 2

The data in Fig. 1 will be input into the computer in Fig. 2.

Integer

DEFINITION:

A whole number (not a fraction or mixed number)

EXAMPLES:

1, 10, 142, 279, etc.

Interpreter

DEFINITIONS:

1. A person who translates
2. A program that takes a statement that is not in machine language, translates it, and then carries it out, before going on to the next statement

EXAMPLE:

Frenchman   Interpreter   Englishman

Inventory

DEFINITION:

A listing of goods or items

EXAMPLE:

An inventory of a soft-drink store might read:
600 bottles of Coke
300 bottles of Pepsi
1,000 bottles of ginger ale
150 bottles of cream soda
250 bottles of root beer
675 bottles of orange soda
760 bottles of cherry soda

Invert

DEFINITION:

To turn over; reverse

EXAMPLES:

1.

    The first house has been *inverted*.

2. $\frac{7}{8}$ $\frac{8}{7}$

    The first fraction has been *inverted*.

Iterate

DEFINITION:

To repeat; especially to repeat a series of steps in a program

EXAMPLE:

1234, 1234, 1234, 1234.

Keypunch

DEFINITION:

A special machine used to record data on cards by punching holes in them

EXAMPLE:

Card

Logic

DEFINITIONS:

1. Reasoning
2. The parts of a computer or of a computer program that appear to involve reasoning

EXAMPLE:

If (X=0) *go to* 25
This statement in a computer program shows a logical or reasoning type of instruction.

Loop

DEFINITIONS:

1. A circle
2. A series of steps that are repeated

EXAMPLES:

1.

2. READ
   STEP 1   ADD
   STEP 2   MULTIPLY
   STEP 3   SUBTRACT
   STEP 4   RETURN TO STEP 1
             (Here you "loop" back to step 1)

Lower Case

DEFINITION:

The usual type on a typewriter produced when you do not use the SHIFT key

EXAMPLE:

SHIFT keys

Typewriter

Lower case is typed when the SHIFT key is not used.

Magnetic Disk

DEFINITION:

A data storage device that uses a magnetically coated cylinder

EXAMPLE:

Symbol

Magnetic Drum

DEFINITION:

A data storage device that uses a reel of magnetically coated tape

EXAMPLE:

Symbol

Magnetic Tape

DEFINITION:

A data storage device that uses a magnetically coated tape

EXAMPLE:

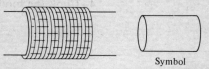
Symbol

Matrix   (Computer)

DEFINITIONS:

1. An arrangement (array) of numbers in a rectangular form
2. Any table can be considered a *matrix*

EXAMPLE:

	A	B	C	D
1	6	5	4	3
2	2	1	0	9
3	8	7	6	5
4	4	3	2	1

Memory

DEFINITION:

A device on which information can be placed (stored)

EXAMPLES:

1. Magnetic tape
2. Magnetic disk
3. Magnetic drum
4. Core

Method

DEFINITION:

A way of doing something

EXAMPLES:

A *method* of preparing a meal is to:

1. Plan a meal.
2. Buy the food.
3. Read the instructions.
4. Cook the food.
5. Serve the food.

Mode

DEFINITION:

A way of operating something

EXAMPLES:

1.

Man operates in a manual *mode* by lifting a package.

260 | **Essential Math, Science, and Computer Terms**

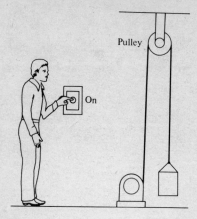

2. This man is operating in an automatic *mode* (pushing a button). When he pushes the button, the pulley lifts the weight.

Output

DEFINITION:

Results that are taken from a computer and sent somewhere else

EXAMPLE:

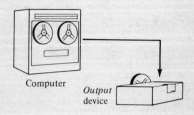

Precedence

DEFINITION:

The order in which things are done

EXAMPLES:

1.

Because the small boy was in line first, he has *precedence* over the man.

2. A car facing a green traffic light has *precedence* over a car facing a red traffic light.

Program

DEFINITION:

The complete set of instructions given to a computer in a special form and sequence so that the computer will perform certain tasks

EXAMPLE:

```
START
INSERT   X=1; Y=2
         Z=X+Y
CALCULATE Z
RETURN TO 1
```

Punch Card

DEFINITION:

A card that has holes punched in it at various places and is used to enter data into a computer

EXAMPLE:

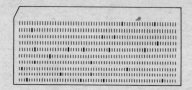

Random Access

DEFINITION:

> A method of storing information in no special order

EXAMPLES:

> Magnetic disks and drums are *random access* devices; magnetic tapes are sequential (in order) devices.

Real Time   (Program)

DEFINITION:

> A type of computer program that continuously reacts to new input data

EXAMPLE:

Airline reservation system:

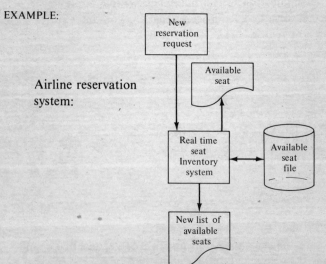

Record

DEFINITION:

1. Something written down for later use
2. One part of an information file, especially a computer file

EXAMPLE:

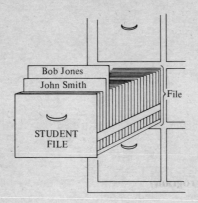

Registers

DEFINITION:

The parts of a computer in which arithmetic can be performed

EXAMPLE:

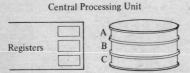

To execute $C = A + B$, A is moved to the register, B is added to that register, and the results are then moved to C.

Remote Job Entry

DEFINITION:

The use of a computer terminal to cause a job to be run on a computer

EXAMPLE:

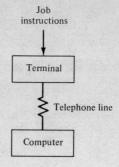

Repetitive

DEFINITION:

Being done over and over again

EXAMPLE:

1, 2; 1, 2; 1, 2; 1, 2; 1, 2; . . . . .

This sequence of the numbers 1 and 2 is *repetitive*.

Retrieval

DEFINITION:

The act of getting some stored material back (such as from a file)

EXAMPLE:

Retrieval of an apple from a box of apples is shown above.

Routine

DEFINITION:

A series of actions that is done often

EXAMPLE:

A computer program for typing letters of admission to a university is a *routine*.

Scan

DEFINITION:

To look over stored information for a special reason

EXAMPLE:

We might *scan* the files of the box to pick out only those files constructed like ⊥.

Schedule

DEFINITION:

A list of events in the order they should occur

EXAMPLE:

A soldier's daily *schedule*:

6 a.m.	Wake Up
1—7:30 a.m.	Eat Breakfast
8 a.m.—noon	Drills
Noon—1 p.m.	Lunch
1—4 p.m.	Classroom Study
4—5 p.m.	Wash Up
5—6 p.m.	Supper
6—10 p.m.	Free Time
10—6 a.m.	Sleep

Sequence

DEFINITION:

The order in which something is done

EXAMPLES:

1. The normal *sequence* of numbers is 0, 1, 2, 3, 4, 5, etc.

2. The normal *sequence* of letters is A, B, C, D, E, F, etc.

Sequential Device

DEFINITION:

A data input/output device from which data are read or into which data are written in order; nothing can be skipped

EXAMPLE:

Magnetic tapes and computer card readers are *sequential devices*. Computer disks and drums are random access devices.

Software

DEFINITION:

The programs, routines, compilers, etc., used in/with computers; i.e., all computer-related materials that are not hardware

EXAMPLE:

A set of instructions (program) written in Fortran to be used in a computer.

Sort

DEFINITION:

Putting data into some desired order

EXAMPLE:

The numbers, letters, and symbols below have been sorted into their own tables:

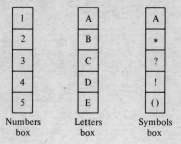

Specification

DEFINITION:

A detailed writing or telling of something in such detail as to define it very clearly

EXAMPLE:

A detailed specification of a box to be used in shipping an item:

The box shall be 3.2" × 4.8" × 5.9"; shall be made of cardboard; can weigh no more than 12 ounces when empty; should be able to withstand a force of 12 pounds; will not crumble when wet and should under normal use last one year.

Statement

DEFINITIONS:

1. A sentence

2. A single line or instruction in a computer program

EXAMPLES:

1. "GO TO 3"

2. STOP

Stochastic

DEFINITIONS:

1. By trial and error (guess)
2. Involving chance or probability
3. Random

EXAMPLE:

Rolling dice is a *stochastic* process.

Storage (Store)

DEFINITION:

A way of putting something away for later use

EXAMPLES:

1.

The man has *storage* space for food.

2. Computer memory is a *storage* device.

Structure

DEFINITIONS:

1. Method of putting something together
2. A house or building

EXAMPLES:

1.  The punch cards represent a *structured* program.

2.  The barn is a *structure*.

Subroutine

DEFINITION:

A set of instructions to carry out a certain job within a larger routine

EXAMPLE:

A program to maintain bank accounts might use *subroutines* to calculate service charges, interest payments, etc.

Terminal

DEFINITIONS:

1. A beginning or ending point

2. A device for entering instructions and data into a computer; it looks something like a typewriter and is connected to the computer by a telephone line

EXAMPLES:

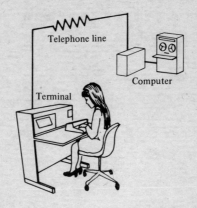

Time Sharing

DEFINITION:

A way of scheduling computer use so that more than one job can be running at the same time

EXAMPLE:

Time	Batch Runs	Time Sharing Run
0	Start Job A	Start Job A
1		Start Job B
2		Continue with Job A
3		Continue with Job B
4	Start Job B	End Job A
5		Continue with Job B
6	End Job A	End Job B
7		
8	End Job B	

Tolerance

DEFINITIONS:

1. The amount that someone or something can endure

2. An allowable difference

EXAMPLES:

1.

The tree has a *tolerance* for only so many ax hits before it falls.

2. If the *tolerance* is .001, any value from .999 to 1.001 will be used as being equal to 1.0.

Transaction

DEFINITION:

A series of related actions, usually related to a single business deal

EXAMPLE:

Booking a flight, setting its schedule and alternate flight, is an example of a *transaction* for one passenger.

Transform

DEFINITION:

To change something into something else

EXAMPLE:

A caterpillar *transforms* itself into a butterfly.

Translator

DEFINITION:

A person (or program) that takes statements in one (computer) language and changes them into another (computer) language

EXAMPLE:

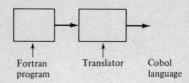

Fortran program   Translator   Cobol language

Truncate

DEFINITION:

To drop off one or more digits in a number

EXAMPLES:

1. 2.1417632 can be *truncated* to 2.14.

2. 3.14159265 can be *truncated* to 3.14.

Upper Case

DEFINITION:

The type on a typewriter produced when you use the SHIFT key; particularly capital letters

EXAMPLE:

Upper case is typed when the SHIFT key is used.

SHIFT keys

Typewriter

Variable

DEFINITION:

A term or symbol that has or may be given different values

EXAMPLE:

The area of any circle is πr^2. r, the radius, is a *variable*. π and 2 are not (they are constants).

Computer Quiz

1. Match the word in column A with the correct figure in column B.

   Column A   *Column B*
   1. Punch card   (A)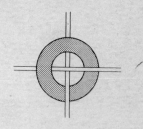

   2. Loop   (B)

   3. Block diagram   (C)

Computer Terms | **275**

Column A   *Column B*

4. Magnetic core   (D)

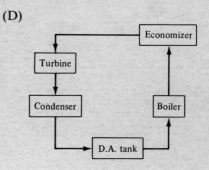

5. Truncate   (E)

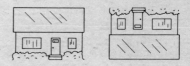

6. Display   (F)

7. Block   (G)

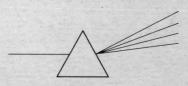

8. Storage   (H)   2.1417632 becomes 2.14

276 | Essential Math, Science, and Computer Terms

Column A   *Column B*

9. Flow chart   (I)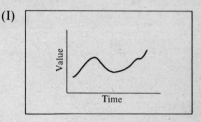

10. Invert   (J)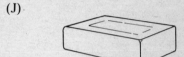

2. Choose the correct word.

   (A) A step-by-step simplified way of getting an answer is an (accumulator) (algorithm).

   (B) The section of an arithmetic unit of a computer is an (accumulator) (algorithm).

   (C) The unit of data storage in a computer usually made up of 8 binary bits is a (byte) (Boolean).

   (D) A way of giving one of two choices, usually "yes" or "no"; "go" or "no go," is a (buffer) (bit).

   (E) A numbering system based on 2's instead of 10's and using only the digits 0 and 1 is a (batch) (binary) system.

   (F) The part of a computer in which arithmetic can be performed is the (retrieval) (register).

   (G) A series of actions that are performed over and over again is a (retrieval) (routine).

Computer Terms | 277

(H) A way of operating something is a (mode) (memory).

(I) The programs, routines, etc., used in a computer are (software) (hardware).

(J) The actual equipment used with computers is (software) (hardware).

3. Pick the equipment from column A that is best described by the words in column B.

Column A *Column B*

(A)  1. Magnetic tape

(B) 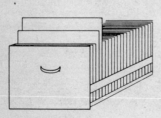 2. Key punch

(C)   3. Console

278 | **Essential Math, Science, and Computer Terms**

Column A   Column B

(D)   4. Magnetic drum

(E)   5. File

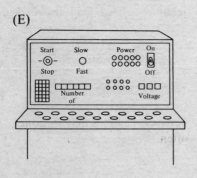

4. List four different types of computer languages.

5. Match the word in column A with the proper definition in column B.

Column A   Column B

(A) Computer language   1. A special program that changes statements from one language into another.

(B) Translator   2. A value that does not change

Column A

(C) Memory

(D) Routine

(E) Subroutine

(F) Inventory

(G) Constant

(H) Variable

(I) Alphanumeric

(J) Logic

Column B

3. A device on which information can be placed

4. Data consisting of letters, symbols, and/or numbers

5. Reasoning

6. A combination of words and phrases by which a person can give instructions to a computer

7. A set of instructions to carry out a certain job within a routine

8. A series of actions that are done often.

9. A listing of goods or items

10. A term or symbol that may be given different values

Answers to Computer Quiz

Question 1

Question	Answer	Question	Answer
1.	(C)	6.	(I)
2.	(F)	7.	(J)
3.	(D)	8.	(B)
4.	(A)	9.	(D)
5.	(H)	10.	(E)

Question 2

Question	Answer	Question	Answer
(A)	algorithm	(F)	register
(B)	accumulator	(G)	routine
(C)	byte	(H)	mode
(D)	bit	(I)	software
(E)	binary	(J)	hardware

Question 3

Question	Answer	Question	Answer
(A)	1	(D)	2
(B)	5	(E)	3
(C)	4		

Question 4

>Fortran
>Basic
>Cobol
>Assembler machine language

Question 5

Question	Answer	Question	Answer
(A)	6	(F)	9
(B)	1	(G)	10
(C)	3	(H)	2
(D)	8	(I)	4
(E)	7	(J)	5